Synthese Library

Studies in Epistemology, Logic, Methodology, and Philosophy of Science

Volume 483

The aim of *Synthese Library* is to provide a forum for the best current work in the methodology and philosophy of science and in epistemology, all broadly understood. A wide variety of different approaches have traditionally been represented in the Library, and every effort is made to maintain this variety, not for its own sake, but because we believe that there are many fruitful and illuminating approaches to the philosophy of science and related disciplines.

Special attention is paid to methodological studies which illustrate the interplay of empirical and philosophical viewpoints and to contributions to the formal (logical, set-theoretical, mathematical, information-theoretical, decision-theoretical, etc.) methodology of empirical sciences. Likewise, the applications of logical methods to epistemology as well as philosophically and methodologically relevant studies in logic are strongly encouraged. The emphasis on logic will be tempered by interest in the psychological, historical, and sociological aspects of science. In addition to monographs *Synthese Library* publishes thematically unified anthologies and edited volumes with a well-defined topical focus inside the aim and scope of the book series. The contributions in the volumes are expected to be focused and structurally organized in accordance with the central theme(s), and should be tied together by an extensive editorial introduction or set of introductions if the volume is divided into parts. An extensive bibliography and index are mandatory.

Michel Ghins

Scientific Realism and Laws of Nature: A Metaphysics of Causal Powers

Springer

Michel Ghins
Institut supérieur de philosophie
Universite Catholique de Louvain
Louvain-la-Neuve, Belgium

ISSN 0166-6991 ISSN 2542-8292 (electronic)
Synthese Library
ISBN 978-3-031-54226-8 ISBN 978-3-031-54227-5 (eBook)
https://doi.org/10.1007/978-3-031-54227-5

This Springer imprint is published by the registered company Springer Nature Switzerland AG
The registered company address is: Gewerbestrasse 11, 6330 Cham, Switzerland

We philosophers have suffered under the burden of Hume for too long.
It is time for us to return to our philosophical home in a metaphysics of substances and powers – the metaphysics of Aristotle, whose yoke is easy and whose burden is light.

Jonathan Jacobs

Metaphysicians should not expect any certainties in their inquiries.

David Armstrong

Preface

What is a scientific theory? Do we have reasons to believe that a theory is true, at least partially? What is a scientific law? What kind of reality makes a law true? Are laws necessary? If they are, in what sense? What are the arguments in favour of a metaphysics of causal powers to ground the necessity of laws? These are the main questions I address in this book. I attempt to answer them within a *moderate empiricist perspective* which emphasises the role of perceptual experience as the main source for justifying our beliefs.

I first offer a defence of a realist interpretation of our best scientific theories. I contend that it is more reasonable to believe in the truth, at least partial and approximate, of scientific theories whose predictive correctness is impressive rather than to adopt a sceptical position. If we accept this realist view, and thus embrace *epistemological scientific realism*, it is natural to inquire about the nature of the realities described by our best theories. To put it briefly, I will develop a realism of facts, which are instantiations of *properties* at some space-time locations. Existing entities such as gases, viruses, molecules, etc. will be construed as instantiations of properties.

I will argue that we have better reasons to believe in the reality of properties which are directly or indirectly observed by implementing an inductive argumentation more in line with the spirit of an empiricist philosophy. Such strategy is distinct from the usual strategy mobilised by realists, which is based on the explanatory virtues of supposedly existing entities. I will thus *not* argue in favour of the existence of gases, viruses, electrons, etc. on the basis that their reality best explains the observations. On the contrary, I will attempt to show that belief in these entities is justified on the basis of inductively and empirically ascertained causal connections between them and the observations.

Furthermore, I will attempt to show that we have strong reasons to accept the existence of some dispositional properties, namely natural causal powers such as solubility or fragility. Although such properties are beyond observational access, and are in this sense metaphysical, we have strong reasons, based on experience, to believe in the existence of causal powers which echo Aristotelian potentialities.

These real powers ground the necessary truth of scientific laws. I do not subscribe to a realist view of laws. Instead, I take laws to be propositions that describe the regular behaviour of natural entities. Moreover, I am sceptical about the existence of essences and natural kinds. Yet, in contrast to David Hume and his numerous successors who promote a conception of an inert and passive nature reduced to a temporal regular succession of facts without real connections between them, I favour a dynamic vision of nature as a whole, set into motion by causal powers. I will thus embrace a specific version of neo-Aristotelian metaphysics.

I attempted to present difficult topics in the clearest possible way so that this book could be used as a textbook for advanced students in the philosophy of science, but could also interest specialists in the philosophy and metaphysics of science as well as philosophers in general. The scientific examples I discuss are standard and relatively simple. I also refrained from getting into the technical details of complex arguments. I even omitted some of them, which may sometimes give an impression of dogmatism, although I am quite aware that numerous objections, as always in philosophy, can be raised on many points. Moreover, some important issues have hardly been touched upon. For example, I don't discuss problems that have been widely studied, such as the demarcation problem between science and pseudoscience and the empirical confirmation of theories. Instead, I preferred to elaborate at greater length on what I consider the most interesting and challenging issues in the philosophy and metaphysics of science today, like the arguments *pro et contra* scientific realism and the grounds of the necessity of laws of nature.

Most of the views defended in this book have already been championed by philosophers who exerted a great influence on me, among whom I could not fail to mention David Armstrong, Stephen Mumford, Stathis Psillos and Bas van Fraassen. Some of my arguments, especially the one for scientific realism, are, to my knowledge, original. But the novelty of this book lies mainly in the defence of views that have not been simultaneously presented in a comprehensive manner so far. I offer here a synthetic and, I hope, coherent presentation of them.

This book grew from the public lessons I gave in August 2011 at the invitation of Professor Eduardo Barra at the *Escola Paraneense de História e Filosofia da Ciência* of the *Universidade Federal do Paraná* in Curitiba, Brazil. These lessons were published under the title *Uma Introdução à metafísica da natureza: Representação, realismo e leis científicas* (Ghins 2013a). I have modified and developed, sometimes considerably, certain arguments. Although I remained faithful to the main views endorsed in my book in Portuguese, I have revised them on several important respects.

The issues discussed in this book have been with me for many years, during which I had the privilege to benefit from many fruitful exchanges with students and colleagues. Among these Evandro Agazzi, Gennaro Auletta, Theodoros Arabatzis, Eduardo Barra, Jim Brown, Otávio Bueno, Marco Buzzoni, Anjan Chakravartty, Valeria Chasova, Silvio Chibeni, Rodrigo Cid, Alberto Cordero, Nadine de Courtenay, Dennis Dieks, Mauro Dorato, Itala D'Ottaviano, Luiz Henrique de A. Dutra, Michael Esfeld, Vincenzo Fano, Laura Felline, Bernard Feltz, Steven French, Alexandre Guay, Gerhard Heinzmann, Marc Leclerc, Giovanni Macchia,

Rafael Martinez, Tim Maudlin, James McAllistair, Angelo Petroni, Claudio Pizzi, Stathis Psillos, Tom Ryckman, Pierre Saint-Germier, Mauricio Suárez, Daniel Tozzini, Giuseppe Tanzella-Nitti, Daniel Vanderveken, Bas van Fraassen and Jure Zovko particularly deserve to be mentioned. Mario Alai, Jean Bricmont, Diego Marconi, Stavros Ioannidis, Demetris Portides, Howard Sankey, Peter Verdée and especially Harvey Robert Brown accepted to read and make much-appreciated comments on parts of this book. I am very grateful to all of them.

I owe a significant intellectual debt to an anonymous referee who read the whole manuscript very carefully and made a great number of detailed and perceptive comments, which have been extremely helpful to clarify some central issues and forestall important objections.

Finally, my greatest gratitude goes to my wife Véronique and our three sons, Léopold, Arthur and Jean-Baptiste, who give me the joy of loving and living in a healthy realism by preventing me from falling in Thales' well.

Louvain-la-Neuve, Belgium Michel Ghins

Introduction

Unlike many philosophers of science today, I will not focus on scientific activity or practice. Of course, science is an activity – who could deny this? – but I will mostly direct my attention to some the main products of scientific activity, namely laws and theories. In the first chapter, I begin by describing the objectifying attitude, which is typical of science, as opposed to the holistic attitude, which characterises ethical and aesthetic approaches. I then propose a conception of theories, called "synthetic", according to which theories are not primarily made either of propositions or models but contain both propositions[1] (such as laws) and models. As I show, models fulfil a double function: they make *true* the laws of the theory, and they *represent* the structures of properties. I discuss at length how models can represent real entities. In doing so, I am largely indebted to Bas van Fraassen's views on representation, but I diverge from him on some significant points. I conclude the first chapter with a brief presentation of what I mean by mechanisms and explanation in the sciences.

The second and third chapters are dedicated to epistemological scientific realism. There, I evaluate the reasons that entitle us to believe that our best theories provide cognitive access to some external realities. I begin by proposing a refutation of the *loss of reality objection* according to which our models are unable to represent real entities external to us. Such refutation is distinct from the one developed by van Fraassen. I then proceed to argue in favour of the classical conception of truth as correspondence between a proposition and the reality that makes it true (if it is true. . .). According to the correspondence conception of truth, for a proposition to be true, it is necessary and sufficient that there is an external reality – its "truthmaker" – which makes it true. This *truthmaking principle* goes against conceptions of truth, which can be considered to be idealist, especially those that hold that the truth of a proposition is based on mental entities, such as *sense data*.

[1] A proposition is the meaning content of a statement. A statement is a linguistic entity. The statements "this is hard" and "ceci est dur" express the same proposition.

To defend epistemological scientific realism, most authors rely on Hilary Putnam's celebrated *no miracle argument*. For Putnam, the predictive successes of our best theories would be *scientifically inexplicable* and then *miraculous* if viruses, gravitational fields, electrons, etc. did not exist. According to him, realism is an overarching scientific hypothesis, which explains the empirical success of science.

First of all, I claim that scientific realism is not a *scientific* hypothesis but a genuinely *philosophical* position. Science is descriptive, whereas philosophy is normative. I oppose the naturalized conception of philosophy – very much in vogue among strict empiricists[2] – which takes philosophy to be a discipline akin to science. In fact, the debate on scientific realism hinges upon the normative *evaluation* of the soundness of the arguments in favour of the existence of certain entities postulated by our best theories such as electrons, genes, etc. but which can only be indirectly observed.

Second, explanationist strategies tacitly assume a pre-established harmony between what we subjectively judge to be a good explanation on the one hand and external reality on the other. I see no reason to subscribe to a form of idealism according to which objective reality should comply with our subjective requirements of what counts as a good explanation.

In the third chapter, I develop an argument for selective epistemological scientific realism, according to which we do have cognitive access to real entities, which I construe as co-instantiated properties. For justifying belief in their existence, those properties must have been observed, but in a broad sense of observation. This view is in accordance with the moderate form of empiricism I favour. If we adopt direct realism, we are entitled to believe in the reality of immediately observed properties, such as shapes or hardness, in the first place. However, it is also legitimate to believe in the reality of properties that are not directly observed, provided that there are observable in principle and that they have been *detected* with the aid of instruments such as the telescope, the microscope, etc., whose reliability has previously been empirically and inductively verified. More generally, I try to show that an argument in favour of belief in the reality of properties that cannot be immediately observed is cogent provided that it relies on empirically ascertained causal connections between these properties and directly observed ones. Whereas the enthusiasts of top-down explanationist strategies evaluate the merits of several competing explanations and conclude that the allegedly best explanation is likely to be true, I implement a bottom-up strategy which starts from what is directly observed.

Thus, I contend that is more reasonable to believe rather than not to believe in the existence of atoms, not because the atomist hypothesis best explains many phenomena, but rather because we can move from the observations up to the causes that produce them, namely atoms, by adopting a bottom-up strategy. As an illustration of this strategy, I propose a reconstruction of Jean Perrin's famous argument for the existence of atoms. Finally, I show that this bottom-up strategy is not vulnerable to two famous objections levelled against scientific realism, namely the

[2]Bas van Fraassen's antinaturalism is a notable exception.

underdetermination of theories by observations and the so-called "pessimistic meta-induction" heralded by Larry Laudan.

For those who support scientific realism, theories do provide selective cognitive access to some realities that are not directly observable. In the fourth chapter, I attempt to elucidate the nature of these realities. For singular propositions – such as <This is hard> – I claim, as David Armstrong does, that their truthmaker is a particular entity that possesses – instantiates – a universal property. The instantiation of a universal property is a *fact*. My realism is therefore a realism of facts, in the sense of instantiated properties. On this, I side with Aristotle, not Plato.

I then introduce the classical distinction between *categorical* and *dispositional* properties. Geometric properties such as sphericity are classical examples of categorical properties. But I do not restrict categorical properties to only geometric and structural properties. I include in the class of categorical properties all observable properties, including the detectable ones, especially those mentioned in scientific theories. As many empiricists emphasised, the mark of the categorical is observability.

Besides categorical properties, we have dispositional properties, which are not observable, as a matter of principle. Dispositions such as solubility, malleability, fragility endow the particular entities which instantiate them with capacities for action. Scientific dispositions will be called "causal powers". Sugar has the power to dissolve in water. Dissolution is what sugar *manifests* when its power of solubility is exercised or activated. When put in water a sugar cube will *necessarily* dissolve if favourable circumstances obtain (if the water is not already saturated with sugar, for example). Scientific laws describe the behaviour of entities when their powers are activated.

On the other hand, the instantiation of a categorical property, such as sphericity, does not depend on some conditions for its manifestation. A spherical particular unconditionally manifests the property of sphericity (be it actually observed or not). Obviously, a property cannot be both categorical and dispositional.

I take ontology to be the branch of philosophy which deals with the nature of entities which constitute the realm of existence or reality. Including categorical properties in our ontology is not that problematic since they can be observed whether directly or indirectly. But this is not the case with powers given that it would be hard to maintain that they are observable. A broken glass can be observed, but the mere *possibility* of its breaking cannot. If we accept the reality of powers, we enter the field of metaphysics. By "metaphysics" I mean the part of ontology which is concerned with entities that are *unobservable* in principle. Empiricist philosophers are in general hostile to any form of metaphysics, which is why they only admit categorical properties in their ontology.

For my part, I try to defend the existence of potentialities or powers by developing an argumentation which takes experience as its starting point. But before doing so, I examine, still in the fourth chapter, the conceptions of laws supported by philosophers who accept categorical properties only in their ontology.

Scientific laws are typically formulated as universal statements such as "All bodies manifest an acceleration proportional to the net force applied" (Newton's second law of dynamics) which purport to describe actual regularities. But what distinguishes this nomological statement from a true universal statement such as "All Italian bars serve espresso"? In other words, what are the grounds of the nomicity of a law? According to David Lewis, a law is a proposition belonging to a scientific theory organised in an axiomatic system which satisfies specific requirements. This view, inspired by David Hume, is referred to as *neoregularist*. According to it, laws describe purely *contingent* regularities. Laws are supposed to be universally true but do not involve any kind of necessity.

Now, it does seem that laws are in some sense necessary, whereas accidentally true statements are not. Nothing seems to prevent an Italian bar from deciding not to serve espresso (at its own risk...) whereas we have a sense that some events are inevitable. For example, experience compels us to believe that applying a net force to a body necessarily modifies its velocity. To take such necessity into account, David Armstrong and others have proposed to identify a law with a relation of necessitation between categorical properties, although this necessity is relative to some logically possible worlds, among which our actual world. Thus, the ontology of our world (as well as the ones of some other possible worlds) includes an actual relation of necessity between the properties of mass, force and acceleration. The nomological proposition that describes this relation is not universal, but *singular*, since it states that specific properties are related by the necessitation relation. But this singular proposition is supposed to imply the truth of Newton's second law, that is, the truth of the universal proposition: <*All* bodies manifest an acceleration proportional to the net force applied>. I will argue that categorical monists, such as Lewis and Armstrong, who refuse to include powers in their ontology, are unable to deliver a satisfactory criterion to distinguish laws from contingently true universal propositions.

Looking for the foundation of the necessity of scientific laws is the privileged gateway to the metaphysics of nature. The fifth – and final – chapter is devoted to a vindication of a neo-Aristotelian metaphysics based on causal powers the existence of which, I submit, can be indirectly ascertained from experience. I then argue that the necessity of scientific laws is grounded on the existence of these natural powers. As a consequence, scientific laws deserve to be properly called *laws of nature*. For example, (ideal) gases possess the categorical properties of pressure, volume and temperature as well as the power to manifest a state of pressure, volume and temperature that satisfies the Boyle-Charles formula, *i.e.* $PV = KT$ (the product of pressure and volume is proportional to temperature). The proposition <Any ideal gas at equilibrium satisfies the formula $PV=KT$> is a law of nature. I carefully distinguish a law of nature from the *description of the manifestation* of a power. Such description typically consists in a mathematical formula such as $PV = KT$. Since it is not logically contradictory to assume that gases, as we know them by their directly observable properties, do not satisfy the Boyle-Charles formula in some other logically possible worlds, the necessity of the ideal gas law is *relative* to the worlds in which the appropriate powers are instantiated. This mode of necessity, that I dub "natural" is more restricted than logical necessity. The truth of a law is necessary according to the natural mode of necessity but remains logically contingent.

This metaphysics of nature borrows a crucial ingredient from Aristotle's metaphysics, namely the existence of powers. In this sense, it is neo-Aristotelian. I present and criticise in some detail one of the most influential current versions of neo-Aristotelian metaphysics, namely the new essentialism of Brian Ellis. However, I try to show that a metaphysics of laws can do without essences and irreducible substantial natural kinds – purportedly denoted by natural kind terms such as "electron". Unlike Aristotle and Ellis, I claim that natural kinds can be reduced, as David Armstrong suggests, to sets of properties and do not enjoy *sui generis* reality. Since, along with causal powers, I admit the existence of categorical properties, I propose a dualist or mixed metaphysics of nature, as Ellis does. However, unlike Ellis, but like Lewis and Armstrong, I defend a broad conception of categorical properties which includes all observable properties mentioned in scientific laws. Thus, nature is made of categorical and dispositional facts, namely the instantiations of categorical properties (or relations) and causal powers. However, the categorical facts the regular succession of which is described by laws are not separated from each other, as they are in the empiricist tradition inherited from David Hume, which sees nature as a "mosaic" of unrelated facts, as David Lewis says. On the contrary, these temporally successive facts are metaphysically connected since they are the manifestations of powers.

I conclude chapter V by examining some difficulties of the mixed metaphysics I vindicate, namely the relation between categorical and dispositional properties and the controversial notion of natural necessity. Finally, in the afterword, I offer some suggestions for possible future research.

Summarising, the present book offers a selective realist interpretation of scientific theories together with a metaphysics of powers which founds the natural necessity of laws of nature, all this within the framework of a moderate empiricist philosophy.

Contents

Chapter 1
What Is a Scientific Theory?

Abstract In this chapter I first present the objectifying attitude, which is characteristic of science, in contrast with the holistic attitude, which is specific to art, friendship and religion. Then, I define models as structures of properties organised by relations, such as nomological formulas like PV=KT. I explain how models can represent scientific objects, namely systems of properties, by endorsing a structural view of representation. According to the model-theoretic view of theories, these are classes of models, whereas for the syntactic view, theories are sets of propositions. I advocate what I call a "synthetic view" according to which theories are constituted of both propositions and models. Models perform a dual role: they can represent objects and also make true some propositions, among which scientific laws. Since scientific theories are supposed to explain phenomena, I contend that the explanatory power of a theory relies on mechanistic models which make causal laws true. A causal law is defined in a syntactic way as a universal proposition which contains a time derivative (or something equivalent to it), which refers to the effect whereas the other terms refer to the cause. Finally, I briefly discuss the issue of determinism.

Scientific theories are not written in a Platonician sky, a world of immutable ideas somehow cognitively accessible to scientists. Theories are the products of scientific practice. This practice, like all human activities, is directed to specific objectives. Defining these objectives is far from being a simple task. To begin with, we must distinguish between the objectives that are *internal* to the practice of science and the *external* motivations that may lead someone to become a professional scientist and to engage in some particular research. These external motivations are not specific to science. They are shared by many other human activities. External motives can be economic, religious, political, military, social and so on. However, scientific practice as such is driven by specific internal aims. Among these, philosophers of science, as well as scientists, highlight the capacity to predict and explain observations. The objectives of science are essentially cognitive. The main purpose of scientific activity is to construct and produce knowledge of a certain kind, namely theories capable of predicting the course of natural and social processes and explaining them by way of their causes.

M. Ghins, *Scientific Realism and Laws of Nature: A Metaphysics of Causal Powers*, Synthese Library 483, https://doi.org/10.1007/978-3-031-54227-5_1

1.1 The Objectifying Attitude

In order to construct theories to predict and explain observations, scientists must adopt a particular attitude or stance towards the world. When looking at concrete things or entities given in perception, that is, phenomena,[1] they must see them as *systems*. A system is set of parts or elements organised by means of relations. When I look at the *Madeleine à la veilleuse* by Georges de la Tour in the Louvre, I am immediately taken by its stupendous beauty. But, if I wish to, I can detach from this state of wonder to direct my attention to certain components of the painting such as the flame of the candle, the table, the left arm of Madeleine as well as the spatial relations in which these elements stand. In doing so, I see the painting as a system, that is, selected parts or elements of the painting that stand in spatial relations.

To perceive systems within observed phenomena, it is necessary to distance oneself from immediate perception as well as from any emotions, whether aesthetic or otherwise, that one might experience. This distancing allows for the selection of certain properties while disregarding others. For example, if an astronomer is interested in celestial phenomena, she may focus on the periods of revolution of the planets rather than their colour or brightness. Similarly, chemists studying a gas might select its properties of pressure, volume and temperature, rather than its colour or smell, as the domain of their scrutiny.

In proceeding in this way, scientists adopt a selective attitude. They first isolate a concrete thing, such as a gas, within the thickness and richness of phenomena. Their second step consists in picking out specific characteristics or properties, such as pressure, volume and temperature, and organize them into a structure using specific relations. Thus, a perceived entity, like a gas, is reduced to a system the elements of which are the properties of pressure, volume and temperature, organised into a structure in a precise manner. Within such approach, scientists identify properties within the phenomena that are considered to be of cognitive interest, and arrange them by means of relations.

By proceeding in this way, scientists construct a scientific *object*, which is a system of properties structured by relationships between them. A system is a structure of properties.[2] For instance, in the case of a gas at constant temperature, the domain of values of pressure, volume, temperature, can be organized using an

[1]The meaning of the term "phenomenon" is ambiguous. Either it means something purely subjective such as a sensation, or a perceived external thing. Bas van Fraassen, who uses the term extensively, is not always clear about this. When I use the word "phenomenon", I mean something perceived or observed, which is external to us. Phenomena are not *sense data*.

[2]The notions of system and structure are closely related. Systems are mostly taken to be real worldly entities which exemplify a structure. In this case, "structure" refers primarily to the organisation of elements. But in mathematics sets of related elements are often referred to as "structures". I will often use the words "structure" and "system" interchangeably since I believe that there can't be a structure unless there are elements – whether they are mathematical or not – in relation, unlike structuralists such as Da Costa and French (2003). For a general definition of system, based on Da Costa's notion of partial structure, see Bresciani Filho and D'Ottaviano (2018).

equality relation, such as "the product of the values of pressure and volume is equal to a constant".[3]

In this way, an entity is reduced to a system or object that is constructed by considering only a limited number of properties and relations between them and omitting many others. In the realm of science, an empirical object is nothing else than a system of properties chosen from among the properties instantiated in the observed entity.[4]

In his book *The Empirical Stance*, Bas van Fraassen (2002) characterises the scientific attitude as "objective distancing". In order to construct a scientific object, the scientist must in the first place adopt a distancing attitude with respect to the entities selected as targets of enquiry. This attitude has nothing to do with spatial distancing. In order to see phenomena as systems, scientists must necessarily make a separation – a caesura – between themselves as human subjects on the one hand, and the thing from which they construct an object-system on the other. Or again, according to Martin Buber (2010), a scientist must separate the *I* from the *That*. In general, a scientific object has the status an object had in some experimental sciences of the seventeenth and eighteenth centuries, that is to say, "an object that is collected, labelled, put in a museum, that is sliced, dissected, solidified, dyed, and put under a microscope." (Wilson 1997, p. 37)[5] When looked at in this way, a thing or entity[6] is reduced to a scientific object that is completely deprived of intrinsic value. After being reduced to an object, an entity becomes available for cognitive purposes without any restrictions or limits. Being devoid of any intrinsic value, a scientific object can be manipulated, altered and even destroyed to serve purely scientific purposes.

Although they can be endowed with some external value by scientists, scientific objects are in themselves value neutral. An entity, when considered exclusively as a scientific object, is deprived of any inner value or dignity. It no longer deserves to be cared for or respected by virtue of any objective characteristic it may possess. Thus, no scientifically objective property can ground the respect due to the life and preservation of human beings, animals, plants etc. The dignity that commands such respect can only be based on values internal to unobjectified entities. These inner values lie beyond the reach of the objectifying attitude, simply because the

[3]Pressure x volume = constant x temperature. This is the Boyle-Charles formula.

[4]Of course, scientists do not always start from experience but also devise theoretical hypotheses and construct possible objects, as it will be seen later.

[5]Quoted by van Fraassen (2002, p.157).

[6]By "thing" or "entity" I mean any kind of existent, be it a substance, a process etc. As it will be developed in Chap. 4, I advocate an ontology of instantiated properties. At the end of the day, any entity is a set of spatiotemporally instantiated properties or relations.

systems that are the objects of scientific research are entirely devoid of any value in themselves.[7]

The objectifying approach is a defining feature of scientific practice. But it can be applied in other fields as well. One can decide to see the *Madeleine à la veilleuse* as composed of parts, study the relations between them, and try to understand the arrangement between colours, shapes, proportions... In fact, adopting an objectifying attitude requires some kind of reversal of our natural attitude towards the world. When we look at entities around us, whether human or not, we spontaneously apprehend them as *singular totalities* and not as systems. If we pause to reflect on this, we realise that the entities that we immediately see are undivided totalities implicitly endowed with intrinsic value, whether positive or negative. A child spontaneously attributes a value to the entities he perceives, whether positive or negative, typically in terms of their intrinsic powers to produce pleasure or displeasure. The same is true for the people – unduly called "primitive" – who apprehend the world as populated by entities inhabited by ancestors or spirits. Some entities, like certain trees in New Caledonia, are directly grasped as undivided units with which it is possible to enter into an individual and personal relationship. Such singular and unique entities possess an intrinsic value that commands the respect due to them. Far from attributing pejorative connotations to such an attitude, we must see it as fundamentally positive, capable of revealing important aspects of the entities of the natural world that surrounds us.[8]

The objectifying attitude typical of science consists in seeing phenomena as systems. It requires a detachment, a bracketing, a suspension of our personal, immediate and spontaneous grasp of the entities given in perception.[9] I call such a suspension of our natural relationship to things the *primary, primordial* or *original abstraction*. Primordial abstraction consists in seeing any entity as an object, that is, a system of *properties*. It is called "abstraction" because it requires *neglecting* or *omitting* most of the properties of a concrete thing given in perception. Even though an object is constructed by means of a process of abstraction, an object is real if its properties and relations are. In in the sciences, the elements in relation are often (but not always[10]) quantifiable properties, such as the pressure and volume of a gas.

---

[7]These considerations have important consequences for the realm of ethics. These will not be discussed in this book. Let me just remark that it is notoriously difficult to articulate the grounds of the dignity of human beings. Certainly, the ability to feel emotions, to think, to want, to love and to be loved, are features that deserve to be considered in attempting to ground human dignity, regardless of the fact that these complex features can also be studied from an objective and scientific point of view.

[8]Such a personal relationship with perceived entities, akin the one we may have with other people, must be very carefully distinguished from the relations, especially the mathematical relations, by which we organise the elements of a system in the objectifying approach.

[9]It seems to me that we are unable to take both an objectifying attitude and what I call below a holistic (emotional, aesthetic, religious etc.) at the same moment. We certainly can pass very quickly from one kind of attitude to the other. The scientist, *qua* scientist, must try to avoid that holistic elements such as emotions etc. impair scientific objectivity.

[10]In some scientific domains, typically in the social sciences, the properties put in relation are often not quantifiable.

These elements are not things in the usual sense of the term, but properties. It is these properties and the relations among them that are studied in the context of the objectifying approach. The choice of properties considered to be relevant is subject to restrictions that follow from the requirement for scientists to introduce a distance, to control their emotions and to refrain from attributing intrinsic values to objects. In adopting some form of asceticism, scientists must strive to curb any personal – subjective or emotional – involvement with the object being studied. Although this ideal is only partially attainable in practice, it is consciously and deliberately pursued by scientists. This is one of the most important aspects of the search for the highest possible degree of objectivity, which would be achieved by the complete elimination of any factor linked to a particular scientist, whether experimenter or theorist.[11] In fact, the scientific object-system cannot be a singular and unique thing but must be reproducible and repeatable. Such a system could in principle be constructed or reconstructed by any scientist, regardless of personality and the context of research.

Logically following original abstraction – although mostly simultaneous with it – the second step inherent in the scientific *démarche* consists in *selecting* in a given phenomenal entity (or set of phenomenal entities) some properties, quantities or magnitudes, which are considered relevant from the perspective of a given discipline: chemistry, sociology, psychology etc. In this way, the field of research is delimited in a sufficiently precise and restrictive way. The choice of specific properties or quantities is the result of what I call the *secondary abstraction*. In this second step, scientists adopt a particular point of view or *perspective* with respect to the targeted phenomena which leads them to neglect or make abstraction of many properties.

Although one can select a multiplicity of properties and structures from the same phenomenon depending on one's interest or point of view, this does not impair belief in their reality.[12] In fact, the term "abstraction" is misleading since in some contexts abstract objects are considered not to be in principle accessible to perception, such as in mathematics. Here, the term "abstraction" indicates that certain properties, including relational properties, of the phenomenon are disregarded or omitted. It is true that the selection of certain properties and relations implies neglecting other ones, but sometimes the properties that are called "abstract" are precisely the properties on which scientists focus their attention within their modelling process, especially when these are mathematized properties. According to the terminology used here, these properties are not abstract but *extracted*[13] from the phenomenal thing. Henceforth, I will understand by secondary abstraction (or simply abstraction when there is no risk of confusion with the primary abstraction) the operation that consists in selecting or extracting from a phenomenal entity some properties and relations while

[11] For an in-depth discussion of objectivity, see Agazzi (2014).

[12] I will discuss at length the epistemological problem of justifying belief in the reality of some properties in the following chapters.

[13] See Portides (2018).

disregarding many others. By this process we construct an object of enquiry, which is real provided its properties and relations are.[14]

Subjective preferences, personal feelings, circumstances etc., play a decisive role in motivating scientists to focus on some objects or scientific disciplines rather than others. Some prefer physics to chemistry, others psychology to demography or economics. But the extracted *properties* themselves can belong to the entities that are the targets of investigation. They can exist independently of individual scientists and the way they are selected or observed. Therein lies the most common meaning of "objectivity". It is imperative in science that the relevant properties be observed or measured by any observer or experimenter, that is to say, by anyone who utilises the appropriate observational or measuring instruments.

A planet is a concrete thing which has an indefinite number of properties, objective or not. But in celestial mechanics a planet is reduced to a quite restricted set of properties such as position, velocity, orbital period... These properties can in principle be observed, directly or indirectly, by anyone, anywhere and at any time. The scientific object is not a unique, singular entity, but a system that can be multiplied and replicated indefinitely. An electron is just a system of instantiated properties: charge, mass etc. each having a precise value and always present simultaneously. These properties form a system: their organisation consists in their simultaneous instantiation or relations among them in some formula. Nothing resembles an electron more than another electron: they all share the same properties. As John Earman once said, there is nothing more boring than an electron: once you have seen one, you have seen them all... This holds for any scientific system-object. But this does not prevent scientific activity from being exciting. Every day new properties are discovered, new objects constructed, and novel theories developed.

On a clear summer night, far from the city and its lights, like Immanuel Kant we all have experienced wonder at the splendour of myriads of stars running across the firmament, a feeling beautifully expressed by Apollinaire's *Chanson du mal-aimé*:

Voie lactée ô sœur lumineuse
Des blancs ruisseaux de Chanaan
Et des corps blancs des amoureuses
Nageurs morts suivrons-nous d'ahan
Ton cours vers d'autres nébuleuses[15]

The difference between the attitude of the scientist and the one of the poet is striking. Instead of dissecting the Milky Way into a system, the poet sees it as a totality in which we can immerse without making any distinction between the observing subject and the object. The same is true in the aesthetic attitude, when

[14] On this, see Agazzi (2014, p. 104)

[15] *Milky Way, O bright sister*
Of the white streams of Canaan
And white bodies of lovers
Shall we, dead swimmers, follow with ahan
Your course towards other nebulae
(My literal translation)

works of art such as paintings, sculptures etc. are perceived as unique, singular totalities with which, instead of maintaining a distance, we try to achieve closeness, even fusion, by entering into the work of art itself to the point of forgetting that we are looking at it. This is the right way of seeing when we look at the *Madeleine à la veilleuse*. Otherwise, we would fail to sense its beauty. I call this attitude *holistic, sapiential* or *contemplative*. We spontaneously adopt it in our friendly and loving relationships with persons. Such contemplative attitude is the one which should be adopted in relation to nature as a whole in order to be able to value and respect it.

However, don't be mistaken in thinking that I undervalue the objectifying attitude with respect to the sapiential attitude. Both play important roles in human life and knowledge. But we must carefully distinguish them. We surely want to avoid falling into a pervasive materialism that would make the scientific approach exclusively and universally cognitively valid, or into a romantic holism that would excessively value the sapiential attitude to the point of despising the scientific attitude.

1.2 Modelling and Representing

The construction of models is an important part of the theoretical activity of scientists. How do they proceed to construct representing models?[16] Representing is an action that can succeed or fail. According to the *structural view* of representation that I will defend, a representation is successful *only if* it involves a transfer of structure from its target (the represented entity) to the representing artefact (a model, or more generally, a representor).[17] This requirement of *structural similarity* – distinct from resemblance – is a *necessary* condition of success. But, as we shall see, it is far from being sufficient.

1.2.1 What Is a Model?

Let us start with a very simple and familiar empirical example drawn from astronomy. In order to construct a model, the initial step consists in isolating and identifying from the bulk of phenomena concrete entities such as relatively fixed bright points called "stars" and moving ones baptised "planets".[18] Individual planets can be

[16] Of course, I propose here a philosophical approach to modelling and theory building. I do not claim that scientists actually consciously work in the way I describe.

[17] A structural conception of representation is defended, with variations, by Suppes (1967, 2002), Da Costa and French (2003), Bartels (2005, 2006) and Chakravartty (2010), among others. It is criticized by Suárez (2003), Contessa (2007), Frigg (2010), Pero and Suárez (2016), and others.

[18] Recall Bertrand Russell's classical distinction between *mention* and *use*. Quotation marks indicate that the term "planet" is mentioned as a word. For example, it can be said that "planet" contains six letters and also that "planet" designates, denotes or refers to certain entities. Quotation marks are not typically used in common parlance when saying that some planet is closer to the sun than some other one, for instance.

identified by means of their colour and brightness. They are given names such as "Mercury", "Venus" or "Mars". Like all concrete things, they have many objective as well as holistic properties. Within the primary abstractive *démarche*, scientists first view the ensemble of planets as an object-system. Then, by secondary abstraction, they select some properties[19] in order to construct specific object-systems, i.e., sets of specific properties. For example, an astronomer can decide to focus on the observed orbital periods, namely the times planets take to return to the same position with respect to the stars. (Positions are attested by observing properties of brightness and distances in relation to stars). In proceeding thus, the astronomer omits mentioning most of the other properties of planets. Model construction is always incomplete.

When observed from the earth the return periods are not constant. This is due to the motion of the earth around the sun. But these return periods are *on average* equal to what we mean today by orbital period, which is the return period of the planet to the same position with respect to the stars, *as seen from the sun*. The duration of a complete revolution of the planet earth around the sun is approximately equal to 365.26 days.

The next step is to construct a scientific object, which is a limited set of chosen properties organized by relations. In the present example, the object is a *perceptual* or *phenomenal system* or *structure*, whose elements are some selected observed properties of the various planets. These properties are the *average* orbital periods observed directly with the naked eye from the earth. These periods are evaluated by observing the positions of the planets relative to the stars over a long period. The values of the planetary orbital periods are then structured using a "smaller than" relation, which is an order relation.

Notice, and this is an important point, that the elements of the perceptual structure are not concrete things but properties. Periods of revolution are properties instantiated in concrete things, namely planets. A specific planet, say Mars, is a concrete thing with respect to which we can adopt a holistic attitude (for example aesthetic or religious) or a scientific attitude. The planets themselves are not components of the constructed perceptual system. Only some of their properties are. A perceptual system is a system of perceived properties and relations selected among those instantiated in the concrete things under consideration.

By constructing an object – which in this case is a specific perceptual system – a scientist performs the first step of the *modelling process*. A *model can be first characterised as a system of properties, that satisfies, that is, makes true, some propositions.*[20] In the empirical example discussed above, the perceptual system makes true propositions that describe the relations between orbital periods. For

[19] I will often use the term "property" to generally refer to monadic and relational properties.

[20] As said in the Introduction a proposition is the meaning content of a statement or a sentence. For example, the distinct statements "The rose is red", "La rose est rouge", "La rosa è rossa" correspond to the same proposition. To refer to a statement, I use quotation marks. To refer to a proposition, I will use square brackets. <The rose is red> refers to the proposition expressed by the above sentences.

example, the perceptual system satisfies the following proposition: <the orbital period of Venus is smaller than that of Mars>. The constructed scientific object is thus a model in the sense just defined, which may be called the "veridical sense" or "alethic sense".[21] Obviously, models in this sense are not propositions.

Of course, scientists cannot rest content with the construction of perceptual systems. More often than not, by direct observation we get rough values that must be made precise by using instruments or measuring devices. In our planetary example, the measurement results are also structured by the relation "smaller than". In this manner, we obtain what is usually called a "data model". Such data model satisfies the propositions that describe relations between the data, namely measurement results. Like the perceptual structure, the data model makes true propositions such as <the orbital period of Venus, equal to 224.7 days, is smaller than that of Mars, equal to 686.9 days>.

In addition to making some propositions true, models can play another important role. *Models are used by scientists to represent.* In the above example, the data model represents part of the sky (the set of planets) *as* a system of orbital properties. The dual role of models is reminiscent of the Roman double-faced god Janus, with one face looking to the past and the other looking to the future.[22] Similarly, models in science have a dual function: they make propositions true and can represent at the same time. But, how and what can models represent?

1.2.2 Photos and Maps

When we talk of representation, we immediately think of classical examples such as photos or maps that may represent a person or a landscape. In the context of our present culture (but not in the context of other cultures, such as precontact tribes in Amazonia), things such as photos or figurative paintings are immediately seen as artifacts that represent. When I look at someone's photo, I implicitly establish a correspondence between properties I select in the photographed person and some parts of the photo. A given coloured area on the photo corresponds to the face, while another area corresponds to the hair of the person who is represented. This kind of correspondence is called a "function" in mathematics.[23] When I look at a coloured piece of paper which I recognise as a photo of someone, I spontaneously associate properties of the person who is represented with elements of the picture, mostly unconsciously. But I can also consciously construct a function that *codifies* a

[21] From the Greek aletheia, which means truth. This sense corresponds to what Alfred Tarski and Patrick Suppes understood by model in mathematical logic, namely a set-theoretical structure satisfying, that is, making true the sentences of a theory (see Suppes 2002, p. 21).

[22] Da Costa and French insist on the dual role played by models (2003, p. 67)

[23] I will understand mathematics as set theory in a very broad sense, not limited to the mathematics used in the quantitative, exact sciences. All scientific disciplines are interested in relations between certain properties and use, sometimes implicitly, set theory.

correspondence between what the photo represents and the photo itself. If each selected property of the person who is represented corresponds to one and only one element of the photo, and vice versa, I have constructed what is called a "bijective function" or "bijection". In this case, the properties selected in what is photographed and in the photo are equal in number.

Moreover, I choose a function that preserves specific relations of a certain kind, such as spatial relationships in this case. What in the photo corresponds to the mouth is located below what corresponds to the nose etc. A bijective function that preserves relations is called an "isomorphism". The relations organise the elements into a structure, a form, as hinted by the Greek etymology of the word "isomorphism"[24]: the person who is represented and her photo share the same structure or form, but only with respect to the construction of a specific correspondence, which is always based on selected elements and relations. Such construction – implicitly taken for granted in some cultural context – disregards some aspects of the photographed person, like three-dimensionality. The photo is a partial and thus incomplete representation of its target.

A function that maps some properties of a person into properties of her photo while preserving selected relations is called a *representative* or *representational function*. The representational function performs a *transfer of structure* and thus institutes a *structural similarity* between the person and the photograph. According to the structural view, some structural similarity between the representor and what is represented is a necessary – but quite insufficient – condition for a representation to be successful.

Let us now look at an unfamiliar quite exotic example shown in the figure below.

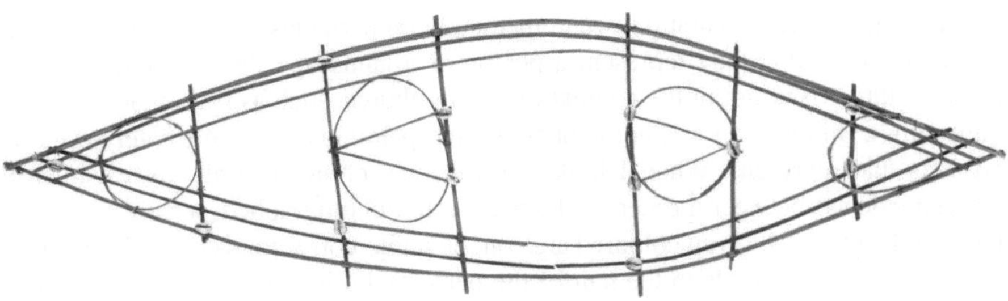

The above artifact is recognised as a representation only in a particular cultural context. Most people are unaware that it was used as a sea chart by Micronesians living in the Marshall Islands.[25] In constructing this artifact, the cartographer intended to convey useful information to a navigator. How did the Marshall Islands

[24]"Μορφη" in Greek translates as "form" and "ισος" as "same". Homomorphisms (one-to-many functions) are more general than isomorphisms. For the sake of simplicity, I will limit the discussion to isomorphisms. Homomorphisms, just as isomorphisms do, can ground structural similarity.

[25]Anthony J. P. Meyer (1995, p. 616, figure 709). This sea chart is in the Linden Museum in Stuttgart. I am grateful to Anthony Meyer and Ulrich Menter, curator of the Linden Museum's Oceanic art collection, for permission to publish this reproduction.

navigators manage to get interesting information from this map? In order to obtain correct, and therefore useful, information, I need to know the code. This code is not included in the map: it is *external* to it. Taken by itself, this artifact does not represent anything at all. Yet it is obviously true that it has internal properties. It is made of wooden sticks tied by knots, contains shells placed at some intersections and other features. When told that it is a maritime chart, we realize that the Micronesians mapped some elements and relations relevant to steering a canoe into elements and relations of the artefact, but we do not know which ones. In order to be able to see and use this artifact as a map, and not look at it as a work of art, I need to know the code, that is, the function – more precisely the isomorphism – that links selected relevant properties and relations of the maritime environment – which is the intended target – to selected properties of the map and selected relations among them.

In order to be able to use this artefact as a helpful instrument for navigation, I need to know that islands are sent to shells. In addition, to detect the presence of a distant island, the Micronesians relied (besides the bearings of stars and other clues) on the observation of interference patterns of swells. Swells are quasi-permanent waves produced by persistent and strong winds, especially the trade-winds near the equator which blow from the east (called "easterlies"). When an island is present, it reflects and refracts parts of those swells (refraction here means the bending of the inshore ends of swells by friction with the island coast). To put it briefly, the presence of an island distorts the swell interference pattern of the open ocean. Such distortion can be observed (by seeing but mostly by the body feeling the waves under the canoe) by experimented navigators up to about a 100 km distance to an island. Swell interference patterns were rarely (to say the least...) taken to be relevant to navigation by Western explorers. However, what is represented by the arrangement of sticks and shells is the distortion of the ocean swell interference pattern by the presence of one or more islands. The map is used to partially represent the maritime environment – the concrete target – *as* a system of swells.

Yet such a map was not used during navigation. In fact, it is a pedagogical and mnemonic tool.[26] Only oral tradition in a specific context could initially reveal how and what it represents. Moreover, the information encoded in this map is quite insufficient to serve as a guide for navigators in Micronesia. Such sea charts are far from being complete since no information from the bearings (azimuths) of stars, among other relevant guiding information, could be gained from them.

But there is more. Suppose I am on a beach on the Ailinglaplap Atoll (where the above stick chart comes from) in the Marshall Islands. To be able to use such a map, I must first locate myself on it, *i.e.* I must know which shell the Ailinglaplap Atoll corresponds to, according to the intention of the maker of the map. I also need to correctly orient the map in relation to my surroundings. This information is not contained in the map! The map itself is "impersonal" in the sense that it can be used by different people at different locations. Location involves both position and

[26] For a detailed explanation of how these maps were used, see David H. Lewis (1994, pp. 224–252) and Ascher (1995).

orientation. If I locate myself in the wrong place on the map, I will be unable to make correct statements such as: if I sail in this direction, I will reach that island. Thus, proper use of the map presupposes the truth of the statement: I am here on the map. Such statement contains indexical terms. "I" and "here" have different denotations depending on who is making the assertion and where the map user is located. Again, just as the code was, the referents or denotations of these indexical terms are determined by information external to the representor.

The examples of the photo and the Micronesian sea chart teach us valuable lessons. Firstly, something never functions by itself as an artifact representing something else. As van Fraassen (2008, p. 23) insists "*There is no representation except in the sense that some things are used, made, or taken, to represent some things as thus or so.*" From the outset, a photo or a map must be *appropriated* by a user for some practical purpose or aim with reference to a particular target. My grandmother's photo is not in itself the photo of my grandmother. It only becomes my grandmother's photo because I *decide* to take such piece of coloured paper as a photo of someone and also specifically the photo of my grandmother. The decision to take the piece of paper as a photo is of course strongly induced by familiar implicit conventions that permeate our cultural context and our knowledge about the making of the photo. Yet this decision is not determined or made compulsory by the context. This holds true also for the decision to make my grandmother the target of the photo.

The example of the Micronesian artifact makes it clear that something does not immediately and obviously present itself as having a particular representing role, except for users in a specific cultural context. In order to represent, something must in the first place be appropriated by a user who determines its target. In themselves, a photograph or an arrangement of sticks, knots and shells do not provide any identification of a possible target. Such identification is achieved by a user who takes some artifact as a picture of a person or as a map that represents a swell pattern. The user's *intention* is paramount: the decision to employ something to represent a chosen target belongs to the user. Different users could use the same artifact to represent different targets. The identification of the target is the result of the user's intentional act which may or may not correspond to the intention of the maker of the artifact. Of course, this intentional act is also external to the target and its representor. As van Fraassen puts it:

> Given this intentionality, it is perhaps not surprising that, in the case of a representation, the relations can change with context of use. The very same object (...) can be used to represent different things in different contexts, and in other contexts do not represent at all. (van Fraassen 2008, p. 27)

The direction or asymmetry of the representation also depends on the intention of the user: a photo or a map represents its target and not the other way around. Since the user's intention and the context play a crucial role in the success of a representation, the relation between a representor R and its target T cannot be reduced to a two-place relation. The representational relationship is a four-place relation: a user U represents T as such and such by means of R in some context C. Given that the user's intentions and the context certainly are external to what the representor and what is represented, the established structural similarity between R and T although necessary is far from being sufficient.

1.2.3 *The Role of True Propositions*

To identify a target, we could simply point to something which is immediately perceptually present. When this is not the case, we must formulate a description that mentions some properties that the target possesses. Identifying the target of a representation is similar to identifying the referent of a word. Referential success can be achieved either in accordance with Kripke's causal theory of reference or (at least partially) true descriptions.[27] In the latter case, the (partial) truth of some propositions is necessary for the identification of the target.

The main point is that identifying the target of a representor is arbitrary, just as a given word can be made to denote anything. Of course, conventions in place in some context determine in practice the denotations of linguistic signs, as a given context does for the targets of some representors. But such external facts do not eliminate the fundamental arbitrariness in attributing a target to a representor. As a consequence, mistargeting understood by Suárez (2003) as applying a representator or a model to "an object that it is not intended for" or to the "wrong target" (Pero and Suárez 2016, p. 74) cannot occur. Indeed, mistargeting is meaningless in the same way as misreferring for a word is, except in the external sense that a word may sometimes be used without complying with the conventions in place in some cultural linguistic context. For example, it could happen that someone who is not proficient in English uses the word "rat" to denote a mouse. Similarly, nothing intrinsic to a Micronesian stick chart prevents it from being appropriated by a user as a map of Louvain-la-Neuve. Then, its intended target would be Louvain-La-Neuve. (I am insisting on this, because it helps to understand why a successful scientific model can be applied to a target, which is very different from its initial target, as we shall see). Obviously, it could be the case that no intended target has been attributed to an entity or an object, as is the case for most entities in the world. Then, such entity does not qualify as a representor in the first place and *a fortiori* cannot misrepresent.

Once a target has been identified, some selected properties, which are intrinsic to the target, are made to correspond to selected properties of the representing artifact, the photo or the map, in order to construct a representational function which preserves some selected relations. Such an operation, which is tantamount to specifying a code, is most often implicit in a given context. When performing a representation, the user always relies (often unconsciously) on a function which performs, as we have seen, a transfer of structure from the target *to* the representing artifact, be it a photo or a map.

---

[27] Theories of reference are discussed in Chap. 3, Sect. 3.7.

It is crucial to realise that the construction of this function is based on the *assumed truth of some propositions* that attribute properties and relations to both the representing artifact (the photo, the sea chart) and its target (a person, a maritime environment) such as <the stick chart contains shells> and < the maritime environment contains islands>. This is the case even if the representation is incomplete or incorrect, that is to say, misrepresents its target. However, such construction does not commit us to a constructivist antirealist position according to which our representations never provide cognitive access to external real things. It is certainly true that in this construction process a thing acquires the status of a representing artifact, but it is also true that it has intrinsic properties. If a photo is used to represent a person it is because people have properties which are put in correlation with properties actually possessed by the photo. This representational action is *successful* only if the target has been identified *and* a transfer of structure from the person to the photo has been carried out by the user, namely someone looking at the photo.

Suppose now that I decide to use the photo of a particular person as a representation of my grandmother. I might do that because at some place on the photo there is a black area above a light area, and my grandmother had black hair. Yet I could have used other identifying clues and other conventions. On the other hand, a photo of my grandmother is considered *correct* or "good" if propositions (which do not have to be uttered or written) about relevant aspects of her physical appearance as well as propositions about some characteristics of the photo are true. The correctness of the photo decisively depends on the isomorphism, which has been established and which conveys to the user some information about my grandmother. If the photo is accurate (in some respects), it is because, among other things, the area and the photo corresponding to the nose is comparatively small and because my grandmother had a small nose, a fact that makes true the proposition <my grandmother had a small nose>.

The very same photo may be intentionally appropriated by someone else to represent my great-aunt. In this case, the photo will *not* represent my great-aunt correctly, i.e. it will *misrepresent* her in some respects because, for example, my great-aunt's nose was larger than my grandmother's. However, it will be a photo of my great-aunt, a successful representor of her based on the truth of propositions (descriptions) identifying her as the intended target.

A representing artifact such as a photograph is neither true nor false. Only statements or propositions can be true or false, such as the proposition <my grandmother's nose was small>. Artifacts and propositions belong to different categories that should not be mixed up. If one speaks of the truth of a representation, it only can be in an indirect or derivative sense. The correctness of an artifact endowed with a representational use rests on facts on which we rely to construct representational functions. Those facts make true some propositions, but an artifact is neither an assertion nor a proposition. For this reason, I will not speak of true or false representations but of correct or incorrect representations.

In what sense can a representor be incorrect or misrepresent? Since the choice of an artifact to represent a particular target is always intentional, the visual resemblance between an artifact and some entity does not determine that the artifact represents something specific, nor *a fortiori* that it represents it correctly in some respects. What is more, it is not true that an artifact must visually resemble its target to be a representation of it. This point is made clear by the example of the Micronesian sea chart. Is it plausible to claim that this chart resembles a maritime environment? If it did, it would immediately be recognised as a sea chart in most contexts. Does a sheet of paper with straight parallel lines and small elliptical patches immediately appear to anyone, in any cultural context, as a music score? Does the score look like the music it represents? Obviously not.

This is why Nelson Goodman could write:

> The most naive view of representation might perhaps be put something like this: "A represents B if and only if A appreciably resembles B". Vestiges of this view, with assorted refinements, persist in most writing on representation. Yet more error could hardly be compressed into so short a formula. (Goodman 1976, pp. 3-4; quoted by van Fraassen 2008, p. 11)

Visual resemblance can certainly play a representational role provided it is encoded in a (usually tacit) representational function that connects identical or nearly identical properties and relations. Obviously, this is implicit in the case of a photograph, which always resembles something or someone by virtue of the very process of its production by means of a particular technical device, such as a camera. Despite of this, a photo can misrepresent. It all depends on the target intentionally chosen by the user. As in the example discussed above, if I decide that the picture of my grandmother is the picture of her sister (who looks very much like her), then I certainly represent her sister. It is the intentional act that decides the target or referent of the representation, and thus its *asymmetry* (the photo represents my great-aunt and not the other way around).[28] However, I could claim that the photo *mis*represents my great-aunt by pointing out that, unlike my grandmother, her nose was large, as we saw. Again, the incorrectness of the photo depends on an actual physical property of my great-aunt.

1.2.4 Some Objections

In this subsection I will focus on two standard objections to the structural view of representation I proposed. According to the first one, the structural view fails to account for cases of non-representation, that is, when no target has been identified and also when the intended target does not exist. Secondly, structural similarity is not a *necessary* condition for representational success in cases such as the chancellor Bismarck being represented as vainglorious, which I will call "symbolic or metaphorical representations".

[28] On the asymmetry issue see Bueno (2010).

If no target has been identified, there is no representation. As seen above, target identification is akin to denotation, since it is arbitrary (anything can represent anything...) just as any entity can be made the referent of a given sign or word. Without target identification, it is obvious that no transfer of structure from a target to a representor can be established by a user. Thus, objecting at this stage that isomorphisms (or, more generally, homomorphisms) have not been instituted or that they are unable to capture the identification of targets is irrelevant.

Granted, in everyday language some objects are often called to represent other things even when no transfer of structure has been established. For example, in board games such as the Game of the goose, it is often said that coloured pieces *represent* the players. But in such instances, no representation takes place, and certainly not in my regimented sense of representation since no structural similarity has been put in place. Actually, it would be more appropriate to say that coloured pieces *stand for* or *denote* the various players.

It must also be emphasised that when an intended target has been properly identified, it need not to be real. Imaginary or ideal targets can be described and theoretically constructed, especially in scientific contexts. Mathematical models are routinely constructed, that may or may not match – approximately – selected properties of real targets. Such activity of model construction plays an important heuristic role in science. Mathematical models or ideal structures such as perfect gases or ideal pendulums are *necessarily* correct (or very nearly so) by construction, in the sense that they exactly (or very nearly so) satisfy the laws[29] of perfect gases and ideal pendulums.[30] Indeed, ideal structures are constructed in such a way that they satisfy some propositions. They are models in the alethic sense. Representation happens only when such models are claimed to partially represent real systems previously identified – independently of theoretical construction – as targets. If a representing ideal model is incorrect in some respects, it can be called a misrepresentation of its target.

A successful and partially correct model for certain targets, can be tentatively applied to very different targets in accordance with the claim that anything can represent anything. In science, such attempts are heuristically guided by analogies supposed to be present between the target (seen as system) of the previous model and the intended target (seen as system) of the tentatively constructed model, as Mary

[29] More precisely, ideal structures satisfy formulas such as PV=KT. The distinction between laws and nomological formulas is introduced in Sect. 1.5.

[30] It has often been pointed out that mathematical structures such as the ideal pendulum are contradictory. To get the law of isochronism of small oscillations ($T = 2\pi (l/g)^{1/2}$, where T is the period of oscillation, l the length of the pendulum and g the gravitational acceleration), one supposes that the angle of oscillation is equal to 0 and thus that the pendulum does not move! Nevertheless, the ideal pendulum with small angles of oscillation nearly satisfies the law of isochronism. And observations of real pendulums approximately agree with the law when oscillations are small...

Hesse (1970) famously contended.[31] Since in the present book I am more interested in the issue of scientific realism and the relation between scientific models and reality, I will not further discuss the heuristic role of models.

The issue of misrepresentation brings us to the second objection, namely when distortion is used to represent the target *as* having properties (or relations) – which I call symbolic or metaphorical – that are not mapped into properties (or relations) of the representor by some isomorphism (or homomorphism).

Let us look more closely at this through one of van Fraassen's examples, namely Spott's caricature of Bismarck (who was chancellor of Germany) as vainglorious.[32] For the caricature to function as a representor *of* Bismarck its target must first be identified. In this example, the referential intention is determined by propositions, which are implicitly supposed to be true by the user and which attribute to Bismarck's face specific physical properties that resemble some features of the caricature. Even if face recognition is achieved without awareness of propositions, it relies on facts such as that Bismarck was bald, wore a moustache etc. which make relevant propositions true. Moreover, a significant part of the structure Bismarck's visage is mapped (by a user in some context) into the structure of part of the caricature. Thus, the target of Spott's caricature is Bismarck.

In the drawing, however, Bismarck's head, chest and feet replace the head, neck and legs of a peacock which displays its tail feathers. Evidently, the drawing was not intended by Spott to be seen as a resembling portrait of Bismarck but as a caricature that represents him *as* vainglorious. How is Spott's aim achieved? As van Fraassen stresses, success in achieving such aim relies on some distortion of Bismarck's physical features. Bismarck's arms are replaced by wings, his chest looks like a long neck etc. If implausibly taken to be a resembling portrait of Bismarck, such representation is obviously incorrect in some respects and is a misrepresentation of him. However, the drawing contains properties and relations – notably the displayed tail feathers – which are typical of peacocks. In our cultural context, peacocks are metaphorically or symbolically associated with vanity. Hence, appropriated by a user in a specific culture, Spott's drawing successfully represents Bismarck as vainglorious, whether he was actually vainglorious or not.[33] Here again, representational success relies on an established isomorphism between parts of Bismarck's body as well as properties of peacocks and properties of the drawing.

Although metaphorical or symbolic representations are not usually used in scientific contexts, distortions are. The representation of a planet-star system as a structure of two mass points is a clear and frequently cited example of distortion.

[31] As Da Costa and French (2003) nicely showed, Hesse's notions of positive, negative, and neutral analogies can be precisely captured by the notions of partial structures and partial isomorphisms.

[32] From Spott (1883). This caricature is mentioned by van Fraassen (2008, p. 14). It can be easily found on the web.

[33] The caricature could have been appropriated to represent a peacock as Bismarckian. Given the context, this would have been a little far-fetched. . .

1.2.5 Conclusion

To conclude this long section, let us recall that representing is an activity or an act, as van Fraassen emphasizes. It involves a user appropriating an artefact to represent an intended target as such and such by stipulating a code and adopting conventions in a certain context. An act is neither true nor false; it can succeed or fail, insofar as the user succeeds or fails to achieve some intended aim. To successfully represent a target, two conditions must be met. First, the user must identify the target. This identification is achieved in an intentional act in the first person: I decide to represent a particular thing (or class of things), recognisable by some of its specific properties, whether it exists or not. This is the reason that mistargeting cannot occur, but only non-representation can, in the same way that a word sign can lack denotation. As a next step, the user must establish, consciously or unconsciously, a transfer of structure from the target to the artifact used to represent it. This requires extracting from the target some properties and relations that are mapped, by an isomorphism (or homomorphism), to selected properties and relations of the artefact. For example, if I decide to use the liquid drop model to represent an atomic nucleus, I must establish a mapping between some properties of the nucleus into properties of the liquid drop model.[34] The success of these two actions (target identification and transfer of structure) presupposes the truth of some propositions about the properties of the target and the representing artifact. I can also successfully represent a person, *albeit* incorrectly in some respects, when using my grandmother's photograph to represent my great-aunt. Since a representing artefact is never true or false, its correctness or incorrectness depends on the code determined by the representative function and – crucially – on some facts which make certain propositions true or false.

Finally, when I represent a given target using some artifact, I always adopt a particular perspective or point of view with respect to some of its properties and relations. By appropriating a piece of coloured paper as a photograph, I represent the chosen target according to some of its visible aspects; in other words, I represent its target *as* a visual system. By appropriating a Micronesian artefact *as* a sea chart, I represent an environment as a system of properties relevant to navigation. In other words, I adopt a certain point of view or a particular perspective on a target by representing it as such or such, that is, according to some highlighted respects. Of course, many other respects are disregarded: my representation is never complete. It is even hugely incomplete because my representation disregards the vast majority of the properties and relations instantiated by the target.

[34]The liquid drop model is discussed by Da Costa & French (2003, pp. 50–51).

1.3 Scientific Modelling

Let us get back to science. Today, many philosophers of science rightly emphasise the prominent role of models in scientific practice. For the "semantic view of theories",[35] theories are primarily sets or classes of models. Such a conception, defended by Patrick Suppes (1967, 2002), Bas van Fraassen (1980, 2008) and Ronald Giere (1988), among others, gradually became prevalent in reaction to the "syntactic conception" according to which theories are sets of statements or propositions. This syntactic view was promoted by the logical positivists of the Vienna Circle in the early twentieth century, such as Rudolf Carnap, Moritz Schlick and Otto Neurath, to mention only a few, and dominated philosophy of science until the late 1960s.[36]

In the *Scientific Image*, Bas van Fraassen endorses the semantic view and claims that "models take centre stage" (1980, p. 44). He of course acknowledges that theories also contain propositions. Nevertheless, a theory is in the first place a set of models that fulfil the dual function of representing phenomena as well as making propositions true. But theories are not maps or photographs. Models in science can be systems of real properties (think of a balls and rods model of the DNA double helix), but also mathematical structures whose mode of reality is controversial and which I qualify as *ideal*.

In the example of the orbital periods of the planets presented above, representing a perceptual system or structure by a data model is achieved by establishing a correspondence between two distinct sets of properties, one which contains directly observed quantities and the other one measurement results. If the domains of the perceptual structure and the data model contain the same number of elements, a one-to-one correspondence between them can be constructed by means of a bijective function F. Such function maps an element of the domain of the perceptual system to an element of the domain of the data model, in such a way that there are no unpaired elements. In the example above, the orbital period of planets observed by the naked eye are mapped to their measured orbital periods.

In addition, we can arrange things so that the correspondence preserves selected relations between the elements of the domains. Two elements a and b connected by a relation R in the perceptual domain are mapped by F to data that stand in a relation R*. In the example we are discussing, R and R* are the same relation "smaller than". Let us call "a" and "b" two observed orbital periods. F(a) and F(b) are the corresponding elements in the data model of a and b, namely their images. If a is smaller than b, then F(a) is smaller than F(b), namely: if aRb, then F(a)R*F(b). The representative function F is an isomorphism. Thus, this set of data can be

[35] The name "semantic" is justified because the semantic conception of theories is primarily interested in models and these, as we saw, can make true some statements and propositions.

[36] See the classic work by Frederick Suppe (1974) for a presentation of the reasons that led to the abandonment of the syntactic conception in favour of the semantic conception of theories.

appropriated by scientists as a model to represent the perceptual or phenomenal structure of orbital periods.

For a data model to represent a perceptual structure, a representative function operating a transfer of structure must have been constructed beforehand by a user. This is the central point of the *structural* conception of representation in science which I advocate. Granted, a mathematical model, such as a data model, does not appear to have the material solidity of a photograph or a map. Models are not propositions. Yet, like propositions, they can be conveyed by means of symbols like written signs on a piece of paper. The relation between a mathematical model and those signs is, I suggest, analogous to the relation between a proposition and a corresponding sentence or propositional sign. Just as the same proposition can be expressed by various sentences in different languages according to distinct conventions, the same model can be conveyed by various material symbols.

We have seen above that the *context* can play a crucial role in determining the representational function for some artifact, seen as a photograph or a map. A given cultural milieu includes a wealth of conventions which determine, most often implicitly, the code assumed by a user in appropriating an artifact as a map, a photo... I also emphasized that the representational relation is complex since it involves four ingredients: the representing artifact, the represented target, the user and the context. However, once the representative function between two structures has been explicitly and clearly specified, as is often the case in science, one can limit the representational relationship to a relationship between two structures of properties, while remaining silent about the other ingredients, which are implicitly taken for granted. I also insisted that I use the word "representation" in a precise, regimented, technical sense, which does not correspond to its usual meaning in some everyday contexts, such as when we say that some piece on a board game represents a player. Also, in attributing property P to an entity S, we commonly say that we *represent* S as instantiating property P. But in this case we are not establishing a representational relationship (in the structural sense) between a property P on the one hand and an entity S on the other, let alone a relationship between an "image", an "idea" or a "representation" in our mind and an external entity.

In the technical structural sense, a successful action of representation requires constructing an isomorphism (or homomorphism) between two structures. A scientist can succeed in representing only by selectively extracting a perceptual structure from a targeted phenomenon. Then, a data structure and an isomorphism between them is constructed in order to finally get a representing data model. Proceeding in this way within the objectifying approach seems to lead to the following counterintuitive consequence: the data model does *not* represent the targeted phenomenon, but only the system of the perceived properties which is extracted from it, that is, a scientific object. At first sight, the representing process seems to imply the unfortunate consequence that scientists loose contact with concrete things. This important issue will be tackled in the next chapter.

Obviously, the elaboration of a scientific theory cannot stop at the construction of a data model, if only because a data model is devoid of explanatory power. The data model only represents the perceptual structure and does not explain *why* the durations of orbital periods increase with the distance to the sun. In order to explain the measurement results, scientists immerge or "embed" the data models into larger, more encompassing structures, called "theoretical structures". What does this *embedding* consist in?

Embedding a data model in a theoretical structure amounts to constructing an isomorphism (or homomorphism) between a substructure of the theoretical structure and the data model. By extension, we can speak of the embedding of perceptual structures, and even of phenomena themselves according to van Fraassen (1980), in theoretical structures, which make true (or nearly so) the laws of the theory. In this broad sense phenomena are embedded in theoretical structures. These theoretical structures are mathematical structures. They are set-theoretical structures capable of representing structures of properties instantiated in an intended target. Furthermore, theories contain structures that cannot in fact be instantiated in reality. As a matter of fact, there are no completely isolated systems of two massive bodies which accurately match the mathematical models of Newtonian mechanics. Real systems are represented by mathematical models only in an approximate way.

In celestial mechanics, the structure of orbital periods can be embedded in the set of models of Newtonian mechanics that deal with two masses in gravitational interaction. As a first approximation, when looking at the orbital motion of a planet such as Mars we focus on Mars and the sun only and disregard the gravitational influence of other planets. Since the mass of the sun represents more than 99% of the total mass of the solar system, such omission is justified when a high degree of precision is not required.[37] Moreover, we take planets and the sun to be point masses. When simplifying assumptions disregard properties which are known to be relevant (such as the presence of more than two gravitationally interacting bodies), this is an aspect of the abstracting process. However, the simplifying assumptions may not conform to some relevant properties of the target (such as treating planets as point masses). In such cases, some relevant properties are modified or distorted. In situations like these, we should speak of *idealisation* as Portides recommends. (Portides 2018) As a consequence of omissions and idealisations, a proposed model can be incorrect in several relevant respects. Scientists are aware of this. This is why they strive to construct more complete and more accurate models by taking into account relevant factors that have been omitted by abstraction and also by modifying or cancelling idealisations. In practice, simpler or more sophisticated models are used in different contexts in function of the degree of correctness which is pursued.

Newtonian theory contains a theoretical substructure isomorphic to the data model of orbital periods. But the availability of an isomorphism does not warrant that the theoretically calculated values of orbital periods conform the data. In addition, we want the theoretical values to be sufficiently close or *adequate* to the measured values. Otherwise, the two-mass-point model would be useless for making correct predictions. A theoretical substructure susceptible to be isomorphic and also

[37] Or when the mathematical tools to deal with more than two masses were not yet available.

adequate to the data model is called an *empirical substructure* (van Fraassen 1980, p. 64). It is empirical because the properties of its domain, such as orbital period values, are measurable. Yet an empirical substructure is also theoretical since it is obtained by calculation in the framework of a theory. Given that its adequacy can be tested empirically, there is no contradiction for an empirical substructure to be both empirical and theoretical. When for any data model belonging to the domain of investigation of a theory, the theory always contains an empirical substructure that adequately, that is, correctly (at least approximately), represents the data model, the theory is said to be "empirically adequate".

Of course, an empirically adequate theory must also allow for the calculation and prediction of *new* measurement results. This is a demanding requirement. Empirical adequacy is not limited to available data but must encompass all possible future measurements. A theory is empirically adequate if and only if it contains empirical substructures that are isomorphic to the set of data models that can be constructed for the set of perceptual structures that fall within its domain of investigation. Obviously, it may happen that new data do not conform to the predictions of the theory. A scientific theory always runs the risk of being falsified: there is no guarantee that future measurement results will always fit the predictions of the theory. This typically occurs when more accurate measuring instruments are devised. If the predictions deviate too much from new measurement results, then the theory can no longer be taken to be empirically adequate.

The process described above can be summarised in the following scheme:

Phenomenon (sky)
@ Abstraction (selection and omission of properties and relations)
Perceptual structure (perceived orbital periods)
↓ Isomorphism
Data model (measured orbital periods)
↓ Isomorphism
Empirical (*and* theoretical) substructure (calculated orbital periods)
∩ Set-theoretical inclusion
Set of theoretical models (two-mass models)
∩ Set-theoretical inclusion
Class of models (theory) of classical point mechanics

It should be clear that this scheme is not supposed to reflect the manner scientists always proceed, let alone the way classical mechanics has been elaborated. Rather, it is meant to be a conceptual epistemological analysis of the connection between theoretical models and observed things. Once the laws of classical mechanics are known, structures that make true those laws, or rather solutions of them, can be constructed and studied independently of experience. Such purely theoretical models could then be applied to some real systems, including systems that do not pertain to the domain for which a theory was originally designed. In other words, a theoretical model can play a heuristic role as I mentioned earlier.

I acknowledge that the presentation above offers a simplified view of the complex process of model construction in science, as the abundant current literature on models testifies. However, the value of simplification lies in its ability to facilitate generalisation. I submit that the proposed framework generally applies to successful modelling in science.

1.4 The Requirements of Scientificity

Having discussed how models represent and before looking at the role of laws, it would not be inappropriate to specify the requirements a theory must satisfy to deserve the title of scientific. Besides empirical adequacy, the virtues most often commended by philosophers of science and scientists are *universality*, *simplicity* and *explanatory power*. Traditionally, strict empiricist philosophers attempted to reduce the conditions that a "good" scientific theory must satisfy to empirical adequacy, formal features or pragmatic benefits. For them, only the observational content of a theory has cognitive value. Accordingly, the cognitive content of a theory exclusively consists in the empirical substructures that correctly represent data models. The other virtues of theories that cannot be grounded on empirical adequacy, such as simplicity, must be shown to be purely formal or pragmatic, that is, devoid of any empirical content. Empirical adequacy certainly is the main requirement for acceptance of a scientific theory. But accepting a theory also implies the commitment to use it in practice to explain and anticipate observations. Moreover, if we are in a position to modify the values of some properties, such as the initial position and velocity of a body, we can control its motion, that is to say, the way in which its properties of position and velocity change with time.

A predictively accurate theory will in principle be preferred to another whose predictions are less precise. But the requirement of accuracy – unlike the notion of accuracy itself which is mathematically well defined – is notoriously difficult to characterise in a precise way. The main reason is that this requirement depends on the context in which the theory is used. In some contexts, facility of use will be favoured over precision. This is a pragmatic criterion. Using classical mechanics or geometric optics rather than general relativity or quantum mechanics in many practical applications within their overlapping empirical domains is quite justified if the degree of accuracy required is not too high. Nevertheless, the philosopher of science – and many scientists as well – will consider that a theory is more empirically adequate, and thus more satisfactory from a cognitive point of view, when its predictive capacity reaches a higher degree of accuracy.

Universality is another highly regarded virtue. A more general theory will be deemed to be cognitively better than a theory whose domain of empirical adequacy is less encompassing. The level of generality of a theory depends on the size of the domain of its empirical adequacy. A theory enjoys a higher level of generality if it contains more models into which more observations – both in quantity and variety – can be embedded. This is why empiricists give prominent value to encompassing theories which unify less general theories. Let me just mention some well-known examples. Newton's mechanics incorporates Kepler's celestial mechanics and Galileo's terrestrial mechanics. The synthetic theory of evolution which integrates Darwin's theory of natural selection and genetics. Einstein's special theory of relativity unifies classical mechanics and classical electromagnetism (to the benefit of the latter since classical mechanics is modified while Maxwell's nineteenth century laws of electromagnetism remain unchanged).

The unifying theory, apart from the fact that it frequently allows for prediction of some measurements with greater precision, often leads to novel observations. Attempts to unify existing theories fosters development of new theories that are empirically adequate to novel, even "miraculous", facts. This happens because more general laws have more solutions that can be satisfied by a larger variety of models that could be applied to represent new observational structures. It is well known that classical mechanics was developed from Newton's *Philosophiae Naturalis Principia Mathematica* first published in 1687.[38] In addition to describing the planetary motions with greater accuracy, Newton also predicted the motions of the Moon, the satellites of other planets and the ebb and flow of the tides. In 1758, by using Newton's laws, Alexis-Claude Clairaut and his collaborators predicted the return of Halley's comet to its perihelion around mid-April 1759. Since the comet passed perihelion on March 13, their prediction was off by 33 days only (Halley's comet average orbital period is about 75 years). (Baum and Sheehan 1997, p. 35) This spectacular prediction played a decisive role in winning over the scientific community to Newton's theory at the expense of rival theories, particularly those of Cartesian inspiration.

However, a scientific theory does not only allow us to predict numerous, varied and precise observational results. It is also supposed to provide an *explanation* of them. With their theories, scientists strive to *explain* the evolution over time of observable properties, such as the position and velocity of a planet for a certain time. Predictive power, grounded on empirical adequacy, makes it possible to gain control of the course of some processes, if we are able to act upon the conditions on which the evolution of these processes depends. For example, we can modify the pressure of a gas by altering its temperature. For processes that are beyond our control, we can sometimes adapt our behaviour in function of our knowledge of their evolution in the future. We cannot – at least not yet – influence tomorrow's weather, but we can adapt our dress according to the weather forecast.

A way of clarifying the notion of *explanation* is by defining it syntactically as a kind of *deduction*. Unsurprisingly, this is the approach favoured by strict empiricist philosophers because such construal of explanation does not require positing the existence of entities such as causes, underlying mechanisms, modal properties (possibility, necessity) the observational status of which is deemed to be problematic. The most famous empiricist conception of explanation in science is the one proposed by Hempel and Oppenheim (1948) which identifies explanation with deduction from universal *laws*. An observation is explained if it is possible to logically *deduce* (often by calculation) the statement which describes this observation from a scientific law and statements describing other observations, which are called the "initial conditions" (or in some cases "boundary conditions"). Thus, Newton's theory explains the position and velocity of a planet at some time by deducing it from laws and initial conditions, namely the position and velocity of the planet observed at another time.

[38] For an English translation see Newton (1971).

Since unification enhances predictive power, it also raises the explanatory power of theories by increasing the quantity of observational statements that can be deduced by means of the laws of the unified theory. For Hempel and Oppenheim, a theory is not a set of models but a set of propositions, in accordance with the syntactic conception of theories. Hempel and Oppenheim's account of scientific explanation – the deductive-nomological (DN) view[39] of explanation – is based not only on scientific laws, but also on a view of theories as primarily organised sets of propositions and not classes of models, contrary to what is advocated by the semantic view. But pragmatic approaches of explanation that similarly avoid resorting to unobservable posits have been developed within the model-theoretic conception of theories by Bas van Fraassen (1980, Chap. 5), among others. In Sect. 1.6 below, I will distance myself from syntactic and pragmatic accounts of scientific explanation and propose a causal mechanistic alternative which nevertheless remains faithful to a moderate form of empiricism.

1.5 The Synthetic Approach to Theories

In this section, I present a conception of theories which, for want of a better term, I shall call *synthetic,* and which attempts to reconcile the semantic and syntactic approaches while preserving the merits of each. According to this synthetic conception, a scientific theory is an organised set of models and propositions, some of which are laws, which assert that some things instantiate properties and structures represented by the models of the theory.

Let us examine this more closely through an example particularly cherished by philosophers of science, namely the kinetic theory of gases. A concrete gas is an entity which has, among its properties, a pressure, a volume and a temperature. These selected properties are directly observable and form the domain of the perceptual structure. At this level, their values can only be roughly estimated. These values stand in some relations. For instance, we can directly observe that when a gas is heated at constant volume its pressure is increased. The observed values are mapped to measured properties of pressure P, volume V and temperature T[40] belonging to the data models in which they are organised by the Boyle-Charles

[39]Also called the "covering law model of explanation". But here, the term "model" doesn't obviously have the same meaning as the one discussed in this chapter.

[40]The concept of temperature has its origin in the sensations of hot and cold. History of science informs us that it has been quite difficult to scientifically define this concept (see van Fraassen 2008, pp. 125–130). The same applies, although to a lesser extent, to pressure. However, someone who has grasped the scientific concept of temperature is able to observe directly *that* two bodies have different temperatures provided the difference is sufficiently large. I develop this important point in Chap. 3.

formula PV=KT (K is Boltzmann's constant), in such a way that the relations of the perceptual structure are preserved. Thus, an isomorphism has been constructed.[41]

According to the kinetic theory, a gas is composed of a very large number of molecules in motion, which perpetually collide in conformity to the laws of classical mechanics. Taking over the scheme presented for planets and applying it to gases, we have:

Phenomenon (gas)
@ Abstraction
Perceptual structure (directly observed properties of pressure, volume, temperature)
↓ Isomorphism
Data model (measured values of pressure P, volume V, temperature T)
↓ Isomorphism
Empirical (and theoretical) substructure (calculated V, P and T)
∩ Set-theoretical inclusion
Theoretical model (V with P and T mathematically defined by means of the average mean velocity of the gas molecules)
∩ Set-theoretical inclusion
Class of models (theory) of statistical point mechanics

The kinetic theory includes the Boyle-Charles ideal gas law. A gas is called "ideal" when its density is not too low, and its temperature and pressure are far from its liquefaction point. These conditions are called "normal conditions".[42] They are satisfied for some gases in our immediate environment, like the air in my office, for example. This law is usually mathematically expressed by the formula PV = KT. But notice that such an equation is not a proposition that can be true or false *in reality*; as it stands, it does not assert anything about what happens or can happen in the world, because in the first place it doesn't mention the kind of concrete things in which the two properties referred to are instantiated. I will call such equations (or similar expressions in non-mathematized disciplines) "nomological formulas". Now, the ideal gas *law*, on the contrary, is a *proposition* which can be formulated in the following way: <in normal conditions, any gas is in a state such that the product of the values of its pressure and volume at equilibrium is equal to a constant multiplied by the value of its temperature > or < in normal conditions, all gases satisfy the formula PV=KT > .

The measured values for pressure, volume and temperature, which are the properties in the domain of the data model, are structured by the equality PV=KT. This equality expresses a relation between *properties*, namely the measured values of pressure, volume, temperature, and organises them into a structure which is a data model. A model of measured properties (the data) satisfies or makes – approximately – true the Boyle-Charles *formula*. Strictly speaking, models don't

[41] I am aware of course that the Boyle-Charles relation was not discovered without resorting to measuring devices. Here, I focus on the representation of a real structure (the perceptual structure) by models (the data model), in accordance with one of my main interests in this book, namely scientific realism.

[42] These are the so-called *ceteris paribus* conditions, to which I shall come back in due course.

make laws true since laws mention concrete things, such as gases, in which the properties referred to in the models are instantiated. Such instantiation is usually implicit, but it is not made explicit in the model as such. As we saw above with the examples of the photograph and the sea chart, the identification of the target is external to the photograph or the sea chart themselves. In the present example, the data model as such does not permit us to identify a specific gas as its target.

In the reconstruction I propose, a concrete gas is not identified or defined as an entity that obeys the nomological formula PV=KT, otherwise the Boyle-Charles law would be an analytic statement. A concrete gas can be identified as something that instantiates *some* pressure, volume, temperature; or, perhaps to make things clearer, by other properties such as fluidity, lack of resistance etc. Then, the proposition that the air in this room (approximately) conforms to Boyle-Charles equality is certainly not an empty statement.[43]

According to the synthetic view, the laws and the models satisfy the nomological formulas are tightly connected. As we just saw with the Boyle-Charles formula, the relations that organise the elements in the data domain are described by a nomological formula. This formula is mentioned in the complete formulation of the law. But whereas models are neither true nor false, a law is a proposition, which could be true or false.

Generally, laws mention concrete entities that behave in accordance with nomological formulas. It is clear that Boyle-Charles's law does not apply to planets, for example. Entities mentioned in laws, such as gases, planets, charged bodies etc. are identified independently of the nomological formulas satisfied by the structures of properties extracted from them. Otherwise, laws would become trivially true propositions.

The ideal gas law is a law of coexistence which unlike the laws of succession does not mention time. The law imposes a constraint on the values of pressure, volume and temperature a gas can instantiate when it is in a state of equilibrium, that is, when its global properties of pressure, volume and temperature are constant. Notice that this law does not provide any *explanation*. It does not explain why the gas exerts a pressure on the walls of the container or why it has some temperature.

An explanation of Boyle-Charles' law is provided by the Maxwell-Boltzmann kinetic theory of gases which describes the microscopic processes that make up the *mechanism* underlying the behaviour of gases. A gas is a large number of randomly moving molecules that collide among themselves and also hit the walls of the container. The pressure on the walls is proportional to the number of collisions per unit area and their force. If the volume decreases, the pressure will increase. If we heat the gas, while keeping its volume constant, the velocities of the molecules will be higher and the collisions on the walls will be stronger, which will also cause a higher pressure. Kinetic theory tells us that the pressure is caused by the collisions of the molecules on the walls of the container and that the temperature increases with the average velocity of the molecules.

[43] Again, these are important issues that will be further discussed in the next chapters.

The motions of molecules conform to the laws of classical mechanics. By relying on those laws, the macroscopic properties of temperature and pressure can be defined by means of the microscopic properties of mass and average velocity of the molecules. The kinetic theory of gases deductively implies the Boyle-Charles law. However, this is not enough to have an explanation. To achieve this aim, the kinetic theory describes an underlying mechanism which grounds the truth of the ideal gas law. Let us now have a closer look at mechanistic explanations.

1.6 Explanation and Causal Mechanism

When attempting to provide a scientific explanation, the first step is to specify what is to be explained. On this, all agree. As van Fraassen rightly insists, to explain is to answer a *why question* (van Fraassen 1980, pp. 126ff). Why do gases follow the Boyle-Charles law? To this question, the kinetic theory provides an answer. One can also ask why the planet Mars is at a specific location now. This is a request for an explanation of an event described by a proposition such as: < Mars is at location x at time t >. The answer given by classical mechanics is well-known. Suppose we know the values of the position and velocity of Mars at some time t_0, which is called "initial". From these initial values we can calculate, using classical mechanics, the position and velocity of Mars at some other time. We are thus able to make both predictions and retrodictions.[44] From propositions that describe the initial conditions at a given time and the laws of classical mechanics, we can deduce propositions that describe the positions and velocities of a planet at another time.

From a purely syntactical perspective, we are then back to the deductive-nomological (DN) view of explanation proposed by Hempel and Oppenheim. Explaining an event consists in deducing a proposition that describes it from propositions about initial conditions and scientific laws. But for the synthetic conception, theories also incorporate classes of models. As a consequence, explaining events is also to embed them in encompassing mathematical structures. By doing so, do we get a good scientific explanation? Strict empiricists do not hesitate in giving a positive answer to that question, because to ask for more would carry us beyond what could be justified on the basis of experience.

However, throughout history many philosophers and scientists have been far from being happy with such a restrictive notion of explanation. Leibnizians and

[44] If we replace *t* by *-t* in the equations, their mathematical form does not change. The laws of mechanics are invariant under time reversal. From this it follows that it is possible to date the past occurrence of an eclipse and a recorded historical event simultaneous to it. In his *History of the Peloponnesian War*, Thucydides relates that, during the Sicilian expedition, the Athenian general Nicias decided to immobilise his fleet during a total lunar eclipse because he thought it to be a bad omen. As a result of this decision, the Syracusans managed to destroy the entire Athenian fleet, a disaster which led to the final victory of Sparta and the end of Athens' hegemony. The lunar eclipse mentioned by Thucydides took place on 27 August 413 BC. (Mozel 1995)

Cartesians were quick to raise objections against Newton's explanation of planetary motions because it appealed to a gravitational force deemed to be mysterious because its mathematical formula does not say anything about what carries gravitation over long distances. In other words, the mathematical formula does not refer to any *mechanism* underlying gravitation. Various mechanisms supposed to explain planetary motions have been offered in the course of history. Ptolemy's crystalline spheres, Descartes' vortices and even angels are well known proposals to that effect.

Classical mechanics however remains completely silent on this issue. And it had to be because the characteristics of an underlying mechanism could not yet be known on the basis of observations. This is the meaning of Newton's famous statement *hypotheses non fingo* ("I make no hypotheses"), which can be found in the *Scholium generale* at the end of the *Principia*. There, Newton warns against any speculation about a carrier of gravitation inaccessible to observation. Such speculations, called "hypotheses"[45] by Newton, are metaphysical. Lacking empirical basis, they have no purchase in science. Surely, the angels of the Scholastics and Descartes' vortices are entities unobservable in principle.

The explanation of the ideal gas law is not just a deduction from the more general laws of classical mechanics. Actually, the kinetic theory delivers an explanation of this law based on an underlying *mechanism*, which consists in colliding moving particles in accordance with the laws of classical mechanics. Today, *new mechanists* such as Wesley Salmon, Stuart Glennan and Peter Machamer claim that a theory is explanatory only if it describes a mechanism responsible for the observable phenomena (Salmon 1984; Glennan 1996, 2000, 2019; Machamer et al. 2000). For them, to have an explanation, it is not sufficient that observational propositions be deducible from the theory. In addition, a genuine explanation must contain premises that describe a mechanism. This is all very well. But the notion of mechanism must then be made precise.

For Galileo and Descartes, a mechanism is a set of geometrically characterised parts in relative motion, such as the standard example of the pendulum clock. The kinetic theory of gases provides an explanatory mechanism in this sense. Indeed, a gas is a set of moving molecules that collide with each other, as billiard balls do. The molecules are spatially distinct entities, as are the parts of a pendulum clock. But whereas the mechanisms of Galileo and Descartes involve the properties of position, velocity and geometric form only, the mechanisms of Newtonian mechanics include masses and forces. However, both collision and gravitational interaction are left unexplained in Newtonian physics. Today, collisions between billiard balls as well as collisions between molecules have been accounted for, first by Maxwell's electromagnetic theory and then by quantum mechanics. Maxwell's theory predicted the existence of electromagnetic waves, which were later associated with a particle in quantum mechanics: the photon. With his theory of gravitation, namely general

[45]Newton uses the term "hypothesis" in a sense distinct from today use. For us, a hypothesis is a supposition that guides research and that could be confirmed or falsified. For Newton, it is a speculation without any connection to experience.

relativity, Einstein was able to predict the existence of gravitational waves which are solutions of his field equations. Gravitational waves have been detected in September 2015: they are tiny perturbations of space that propagate at the speed of light (Abbott et al. 2016). They are now associated with a particle, the graviton. Such historical developments manifest a progressive broadening of the notion of a mechanism, as I will now try to show.

Stuart Glennan states that "a mechanism for a behaviour is a complex *system* that produces that behaviour through the interaction of a number of parts and where the interaction between the parts can be characterised by direct and invariant generalisations about the changes that take place." (Glennan 2000, p. S344).

A mechanism is thus a system of spatially distinct components that interact with each other. The "direct and invariant generalisations" are the laws that describe the evolution of the system over time, i.e., the changing relations between its parts. In general, according to Glennan, these relations are spatial, but they need not be. In line with what I said before, I identify these parts with sets of instantiated properties, some of which are referred to by the laws. (Ghins 2019, p. 105) In doing so, the notion of mechanism is considerably broadened. The constituent parts of mechanisms become properties that are not necessarily spatial or geometrical. Mechanisms in the classical sense, are in fact systems of parts which are clusters of properties, such as position, velocity, mass, force etc. The behaviour of a whole entity such as a clock, a planetary system etc. is the result of the changing relations between their parts. These changing relations are described by laws. For science, a mechanism, in the broad sense in which I understand it, is nothing else than a system of clusters of quantifiable properties (i.e., a system of parts) that stand in mathematical relations. I will develop this enlarged conception of mechanisms and mechanistic explanation in the next section.

1.7 The Extended Mechanistic Explanation

The founding fathers of modern science, Galileo, Descartes and many others, believed that physics, and indeed science as a whole, could be based on a small number of quantifiable properties, the "primary" qualities, namely impenetrability and geometrical properties. They also believed that all other qualities, called "secondary", such as colours, smells and sounds, were reducible to the primary qualities. This dream has not been fulfilled, but it survives in an attenuated form in the objectifying approach, which consists in considering, by abstraction, all entities as systems of properties, which can be represented by models. Of course, these properties do not have to be mechanical properties in the classical sense, but the parts of a mechanism are most often spatially distinct and the changing relations between these parts amount to the functioning of the mechanism. In this way, the notion of "mechanism" is considerably enlarged, while retaining some resemblance with the Galilean-Cartesian conception.

Very interestingly, we can construct a representational function between a mechanism and a structure of specific relations, that is, a model. This is no surprise considering that a mechanism is a set of properties structured by relations. Because of this, models play a prominent explanatory role in science. Thus, to repeat, deductibility of observational propositions from laws is a necessary condition for an explanation of phenomena. But it is not sufficient. Some mechanisms which produce phenomena must also be represented by some models of the theory. Together, deduction and representation are necessary and sufficient conditions to get a satisfactory explanation of some phenomena. This is a welcome result, quite in accordance with the synthetic view of theories which I support.

But not all relations between the components of a system are qualified to make a system a mechanism. As Glennan says, the changing relations between parts are interactions that are described by laws. For him, mechanisms ground causal relations between events. I agree with this ontological claim. But I contend that the laws, namely the propositions that describe the behaviour of a mechanism, must be known as causal in the first place in order to avoid circularity since the word "interaction" has obvious causal overtones.[46] What then is a causal law? For a law to be causal it must satisfy, I suggest, the following sufficient and necessary condition:

> A causal law contains, in its formulation, a temporal derivative, which refers to the effect, whereas the cause (or causes) corresponds to the other terms in the mathematical expression of the law. (see Blondeau and Ghins 2012, pp. 383–384)

More precisely, the nomological formula included in the formulation of a causal law must include a time derivative. Such a way of identifying causes and effects fits our intuition that a cause is what makes a difference in the course of events. The cause is what makes a property to change. The property change is the effect of the cause. In appropriate circumstances, a causal law describes a sufficient and necessary relation between the effect and its cause: if the effect is present, so is its cause, and conversely. In disciplines that are mathematised, the effect is referred to by a time derivative.[47] In Newton's second law for example, a force is the cause of instantaneous acceleration, which is denoted by a time derivative of velocity.

I thus propose the following definition of a mechanism:

> A mechanism for a behaviour is a complex system of spatially distinct parts (i.e., a system of clusters of properties) whose interactions, described by causal laws, produce and explain its behaviour, that is, the changes of some of its global properties with time.

The global pressure of a gas at constant volume increases when heated. Such behaviour is produced and explained by a higher mean velocity of its molecules whose motions are described by the causal laws of classical mechanics, such as Newton's second law. We thus have a mechanistic explanation of the time variation of the gas pressure. Similarly, together with the mathematical expression of the

[46] For a critique of Glennan's view on causality, see Ghins 2019.

[47] This criterion can be extended to laws which are not mathematical, provided they mention a time valuation of some properties.

gravitational force, the causal laws of classical mechanics allow us to deduce the laws of planetary motions (Kepler's laws) and the positions and velocities of planets at any time from some known initial conditions. This provides a *mechanistic explanation* of the *motions* of the planets. However, classical mechanics does *not* provide a mechanistic explanation of the cause itself, that is, gravitation. Such an explanation of gravitation is delivered by general relativity, as mentioned previously. Although I proposed a broad definition of mechanism by admitting that its parts consist of quantifiable properties whose evolution is governed by laws, I imposed a significant restriction on the form of the laws that describe the evolution of these parts over time. For a system to qualify as a mechanism, the relations between its parts (clusters of properties) cannot be described by any kind of laws.

Explanation is an epistemic notion: it is relative to our knowledge. Causality on the other hand, in the sense I understand it, is ontological.[48] Causes are real entities, namely instantiation of properties. In answering a request for explanation, we try to tell why a particular *event* (or *sequences of events*) occurred. As we saw, strict empiricists answer that the explanation of an event boils down to its correct prediction in line with the deductive-nomological scheme. The mechanist, on the other hand, strives to uncover the system workings which underlie correct predictions. The challenge for the mechanist is to find the real mechanism whose behaviour produces the phenomena as we observe them. Such mechanism can be represented by a model. The domain of the model contains elements that correspond to the components of the mechanism; these components are structured by causal nomological formulas. The mechanist philosopher is a scientific realist. For the scientific realist, we have good reasons, grounded on experience, to believe that some mechanisms are real, as I will attempt to show in Chaps. 2 and 3. Since causes are taken to be real, a mechanistic explanation is not epistemic only; it is also committed to existing entities, namely real instantiated properties.

1.8 Explanation and Determinism

To conclude this chapter, let me say a few words on *deterministic* explanation and the status of probability in connection with the issue of realism. A deterministic explanation consists in showing why a specific event occurs *rather than another*; for example, why Mars is at some location at a given time, rather than elsewhere. This kind of explanation is based on deterministic laws. A deterministic law contains a mathematical formula that allows us to calculate, in an unambiguous way, the value of a property at any time from its value at any other time. For example, the laws of Newtonian mechanics permit us to determine by calculation the values of position and velocity of a planet at any other time, past or future from the values of position

[48] It is significant that Wesley Salmon called "ontic" the mechanistic explanation he advocates (Salmon 1984, p. 19).

and velocity of a planet at an initial time (provided the influence of other planets and satellites is considered to be small in relation to degree of precision pursued and can thus be disregarded, since no deterministic law for three bodies in gravitational interaction is available).

It is essential to realise that the availability of a deterministic law is *not* sufficient for predicting the future state of a planet. We also need to know its state at some initial time, for example now, with accuracy. Moreover, we must be able to perform the calculation. *If* we knew the exact values of the initial properties which characterise the state of a system and the deterministic laws that describe its behaviour in time and, on top of this, *if* we had the required mathematical capacity, then we *would* be able to calculate with precision the values of these properties at some other time, past or future. This is what Laplace famously said in his *Philosophical essay on probabilities*:

> We ought then to consider the present state of the universe as the effect of its previous state and as the cause of that which is to follow. An intelligence that, at a given instant, *could* comprehend all the forces by which nature is animated and the respective situation of the beings that make it up, if moreover it *were* vast enough to submit these data to analysis, *would* encompass in the same formula the movements of the greatest bodies of the universe and those of the lightest atoms. For such an intelligence nothing *would* be uncertain, and the future, like the past, *would* be open to its eyes.[49] (Laplace 1995, p. 2)

Such hypothetical intelligence or "demon", envisioned by Laplace, would know with perfect accuracy the initial states of all the components of the universe and all its supposedly deterministic dynamic laws. Moreover, such intelligence is endowed with unlimited computing power. It would then be able calculate the states of all the components of the universe at any time. Even if the universe is such as imagined by Laplace, no human could ever dream to acquire the abilities of this omniscient demon. As Laplace says:

> All these efforts in the search for truth tend to lead the mind continually towards the intelligence we have just mentioned, although it will always remain infinitely distant from this intelligence. (Laplace *Ibid.*, p. 2)

To begin, we are unable to know the initial states with perfect accuracy. Even if deterministic laws and the required computing capacity were available, the lack of precise knowledge of initial states would prevent us from determining a future (or past) state with sufficient precision, especially in the distant future (or past). Thus, even if we lived in a Laplacian universe, we could never have any *certainty* about the evolution of things. Although Newtonian mechanics sets no limits to the accuracy of measurement of initial states, in practice these can never be exactly known. As a consequence, we *cannot* predict the value of a property with certainty. We can only assign a *probability* that this property will have a specific value or that a measurement will give some value of this property. *Pace* Prigogine, uncertainty was

---

[49] Italics are mine.

already present in physics before the advent of chaos theory[50] and quantum mechanics. We did not have to wait for these theories to reach the "end of certainty" (Prigogine and Stengers 1997).

However, there is little doubt that Laplace was a realist and also believed that our universe is deterministic. Interestingly, he didn't justify this belief on the basis of the state of science of his time,[51] but on an *a priori* principle, namely the principle of sufficient reason.

> The connexion between present and preceding events is based on the evident principle that a thing cannot come into existence without there being a cause to produce it. This axiom [is] known as the *principle of sufficient reason* (...) (Laplace 1995, p. 2)

The position and velocity of a body exist independently of performing measurements or not. But in order to know the initial properties, we must measure them. In some circumstances, such as the flip of a coin, we are unable to exactly know its initial position, velocity, rotation, resistance of the air etc. Yet we can assign a certain probability to the outcome of a flip. By convention, probabilities take values between zero and one. An event that we believe to have no chance of occurring is assigned a probability 0. We attribute a probability 1 to an event when we are certain that it will happen. Assigning a probability ½ to an event means that we believe that it has a 50% chance of occurring. For example, when flipping a coin, we are ignorant of its initial state and other surrounding circumstances, but we estimate that the coin has a ½ probability of landing heads (or tails).

How do we get this estimated ½ probability? We know from experience that if we flip a usual coin a large number of times, the outcomes of heads and tails will be approximately equal. In other words, the ratio of favourable cases, heads say, to the number of possible cases, heads or tails, is roughly equal to ½. The *relative frequencies* of heads and tails are nearly equal. Then, by induction we infer that these relative frequencies will remain stable in the future. On this basis, we estimate that if a coin is flipped *only once* the probability of heads (or tails) is ½.

For Laplace, a real external system is in a definite state independently of our knowledge of it. Probabilities reflect our partial ignorance of initial states. Thus, probabilities do not exist in reality but only in our minds. In this sense, they are subjective or, more precisely, epistemic, that is, relative to our partial knowledge of real situations. This, however, does not prevent humans from intersubjectively

---

[50] A system is chaotic when its evolution is very sensitive to a small modification of its initial state. It is then impossible to predict its evolution after some period of time, which can be very short, although chaotic systems obey classical deterministic laws.

[51] Neither could he, given that there are no deterministic solutions in classical mechanics for three bodies gravitationally interacting. Today, methods of approximation are available. Illimited computing power could allow us to get predictions with any degree of precision, but no strictly exact ones. Needless to say, things get much more complicated when more than three bodies are present...

agreeing on the probability of a coin landing heads or on the probabilities delivered by quantum mechanics. Even if for subjectivists[52] a probability is *not* a relative frequency, its value is estimated from observed relative frequencies and other available information (such as symmetry or bias). In this sense, probabilities are objective but not real. What is more, our partial ignorance does not preclude the reality of some underlying mechanism responsible for the occurrence of events such as a coin landing on heads.

Many people get surprised when told that the fundamental equation of quantum mechanics is deterministic. In fact, the Schrödinger equation, which describes the evolution of the state function (or wave function, usually called Ψ) of a system is deterministic. Moreover, the Schrödinger equation is causal because it contains the temporal derivation of the state function, which is the effect, whereas its other terms refer to its causes. In solving this equation, we obtain deterministic equations for the evolution of the quantum states (distinct from the classical states)[53] of a system. It is indeed possible to calculate a determinate value of the state function at a later time from its value at an earlier time. Despite of this, the knowledge of the state function at a given time only allows us to predict the occurrence of classical states (position and velocity, for example) with some (lower than one) probability. A solution to Schrödinger's equation, namely the evolution of the state function in time, ascribes a probability value to the occurrence of each possible measurement result. For example, it is impossible to determine when a given sample of radioactive substance such as radium will emit an alpha particle (made of two protons and two neutrons). However, quantum mechanics allows for a deterministic calculation of the *probability* that an alpha particle will be emitted within some time interval.

The events dealt with in quantum mechanics are different from the ones described by classical mechanics. While in classical mechanics it is *in principle* possible to provide a deterministic causal mechanistic explanation and prediction of the outcome of the flipping of a coin, standard quantum theory does not describe a deterministic mechanism for producing specific events. It "only" (but this is a remarkable achievement!) ascribes specific probabilities to them. In many situations, quantum mechanics does not make us capable to predict and explain why an event occurs rather than another. We are unable to explain why a sample of radium emitted an alpha particle at some instant of time rather than another.

At the beginning of the twentieth century, when quantum mechanics was born, many physicists were most dissatisfied with such a situation. Albert Einstein, among others, claimed that quantum mechanics is an incomplete theory. In other words, the state function Ψ does not provide a complete description of the states of a system, which would allow making precise predictions of single events. To complete

[52]Often called *Bayesians*. When new information becomes available, probabilities are readjusted in accordance with the Bayes formula of the formal probability calculus. For an accessible presentation of the subjectivist interpretation of probability, see Bricmont (2022).

[53]A quantum state is often the sum (superposition) of classical states, each of them having a probability of occurrence when a measurement is made.

quantum theory we should add some properties, for example position in some cases, to the description of the quantum state of a system. The values of these additional properties, if known, would provide complete descriptions of initial states which would permit one to predict and causally explain individual events. These additional properties are called "hidden variables". As early as 1927, Louis de Broglie proposed a deterministic theory in which the hidden variables are the positions of the particles. This theory was later independently developed by David Bohm in 1952.[54] It is still discussed today but remains quite controversial.

Be it as it may, it seems to me that deterministic causal explanations are still a very desirable aim, given the higher level of understanding or comprehension they provide. True, the vast majority of physicists has now gradually become accustomed, if not resigned, to what I consider to be an explanatory vacuum. Today most scientists and philosophers claim that probabilities in quantum mechanics are "irreducible", "intrinsic" or even "real", and that it is futile to look for a deterministic causal mechanism underlying the production of measurement outcomes. This issue is still hotly debated nowadays, mostly among philosophers. I don't want to take sides on this debate. But it seems to me unjustified to assert – as some philosophers and physicists sometimes do in a peremptory way – that it is impossible to develop a quantum theory that would describe a deterministic mechanism for the production of measurement results.

In fact, such a theory exists. It is none other than the de Broglie-Bohm theory, which – at least so far – makes the same empirical predictions as standard quantum theory. The de Broglie-Bohm theory is certainly not free of difficulties,[55] but this also holds true for standard quantum mechanics.[56] Therefore, the proponents of irreducible probabilities and radical indeterminism in nature cannot rely on the current state of quantum mechanics to argue for their position. At any rate, science does not allow us to decide whether the world is deterministic or not. This –

[54] See Jean Bricmont's book (2016, p. 16) and my review (Ghins 2018). Contrary to what many believe, hidden variables theorems do not exclude the de Broglie-Bohm theory. For this theory, particles such as electrons have real positions and trajectories, independently of their measurements. When a measuring apparatus interacts with an electron its state is modified, and the measurement results are consistent so far with the probabilistic predictions of standard quantum mechanics. Admittedly, quantum mechanics sets a limit on the degree of accuracy of a simultaneous measurement of the position and velocity (linear momentum, in fact) of the electron in most cases.

[55] See for example Cushing, J. T. et al. (eds.) (1996).

[56] A well-known difficulty in standard quantum mechanics is the so-called "measurement problem". A measurement consists in the projection of the state function onto one of its components, which corresponds to a measured classical state. This is called the *collapse of the state function*, which is not described by the Schrödinger equation. A solution of Schrödinger's equation describes a superposition of classical states, each of which is ascribed some probability value. The measuring process selects one of them, as in the example of Schrödinger's cat (Bricmont 2016, pp. 38–39). The collapse of the state function is a blind spot in standard quantum mechanics, like collision in classical mechanics before the advent of electromagnetism. Everett's many-worlds theory and the de Broglie-Bohm theory provide solutions to the measurement problem, but they remain a matter of debate. While Brown and Wallace (2005) defend the former, Bricmont (2016) endorses the latter.

philosophical – issue remains largely open. Indeed, even if we find deterministic explanations more satisfactory, this is certainly not a convincing reason to accept Laplace's metaphysical thesis that nature as a whole is deterministic and regulated by deterministic causal laws. Our preferences for a kind of explanation does not entitle us to draw conclusions on the nature of the external world.

Let us conclude this section with a few succinct remarks on the status of probability. It is clear that we do not observe probabilities. We just observe relative frequencies of events, such as the proportion of heads versus the total number of coin flips. According to the subjectivist interpretation of probability which I favour, we assign a probability to a single event on the basis of previously observed frequencies and other information. When we estimate a probability from observed symmetries, such as those of a die, we in fact rely on the previous observation of relative frequencies. Since we have repeatedly observed that a die lands on some specific side with a frequency roughly equal to 1/6, we assign a 1/6 probability to a given outcome when we roll an object exhibiting the same symmetries, other things supposed to be equal (no bias) according to available information.

Empiricists such as Hans Reichenbach (1949) and Richard von Mises (1957) have proposed to identify the probability of an event to the limit of its relative frequency, when the number of cases, e. g. the number of landings of a coin, approaches to infinity. Limits to infinity are mathematically well defined, but they are of course devoid of empirical meaning since only finite sequences of events are observable.

Since probabilities are unobservable, I will adopt an agnostic attitude as regards their existence.[57] But probabilities certainly are not disconnected from experience. The probabilities we assign to events are reliable guides for prediction. Moreover, probability values which enjoy a high degree of consensus can be considered to be objective, in an intersubjective sense. Nevertheless, since probabilities are beyond the reach of observation we have no empirical reason to believe that they are real.

1.9 Conclusion

The scientific *démarche* is characterised by the adoption of an objectifying attitude towards phenomena. By taking an objectifying attitude or stance, a scientist decides to see things as systems of properties, that is, as objects. This stance is distinct from the holistic or contemplative attitude in which an entity, such as a person or a work of art, is seen as a singular totality. It certainly is possible to take an objectifying attitude with respect to a person or work of art, but it is then considered as an object deprived of any value in itself. When scientists get interested in phenomena, observed things, they see them as structures of perceived properties which they can represent by

[57] In my opinion this attitude is recommended for all interpretations of probability. For a discussion of the different interpretations, their respective merits and difficulties, see Hayek (2019).

models. In order to correctly represent a perceptual structure, a scientist constructs a representative function, an isomorphism (or homomorphism), between the perceptual structure and a data model whose domain contains measured values of properties such as position, pressure etc. The success of a representation depends on facts which make true propositions which allow us to identify the intended target. Moreover, models make nomological formulas true, in particular the ones which are mentioned in causal laws, which are essential for a theory to have explanatory power. Causal laws describe the temporal succession of property structures instantiated in sets of things (referred to, for instance, as "planets", "gases" etc.) at different times.

According to the synthetic conception of theories, a scientific theory is a class of models that satisfy (sometimes approximately) mathematical nomological formulas (in a broad sense of "mathematical") and laws which state that some specific things (such as planets, gases etc.), if they exist, will instantiate the structures represented by the models of the theory. To deserve the title of scientific, a theory must contain empirical substructures, by which we can correctly represent structures of results obtained by measuring operations on concrete entities. Data models are structures of measurements results. If the data do not conform to the predictions of the theory, then the theory is not empirically adequate.

Moreover, a theory must be capable not only to predict measurement results as accurately as possible, but also to provide a causal explanation of them. The explanatory power of a theory rests on the causal laws which make possible to construct – through calculation – its empirical substructures. A causal law describes the behaviour of a mechanism, not in the restricted sense of the mechanistic philosophy of the 17th and 18th centuries, but in a broad sense according to which a mechanism is nothing other than a system of quantified properties related by causal connections described by causal laws. To be explanatory a theory must contain models able to represent causal mechanisms.

According to this way of construing mechanisms, causal laws together with mechanistic models ground the explanatory power of a theory. In my opinion, only deterministic causal laws provide a truly satisfactory explanation of the behaviour of a system. Indeed, only in this case we are able to explain why a specific event occurred, rather than another one, in particular instances. Probabilistic laws certainly have great practical use since they can predict with relative frequencies great accuracy, but they are unable to predict, let alone explain, the occurrence of single events, especially when their probability is low.

Since the explanatory capacity of theories is grounded on models and laws, we are now confronted to the task of pondering the reasons that may legitimate our belief in their correctness and truth. We must then tackle the challenging issue scientific realism and evaluate the arguments adduced for believing in the truth, at least partial and approximate, of our best theories. This will be the subject of the next two chapters.

Chapter 2
Scientific Realism: What Is at Stake?

Abstract This chapter begins by discussing what Bas van Fraassen called "the loss of reality objection". Since models are constructed by our action of extracting properties from the phenomena, do we lose contact with reality in this process? While van Fraassen proposes a pragmatic dissolution of this problem, I advocate for a solution grounded on the truth of propositions, which ensure the connection of some models with external reality. Then, I argue in favour of a correspondence view – not *theory* – of truth according to which true propositions are satisfied by external truthmakers. I proceed by showing that the philosophical issue of epistemological scientific realism does not hinge on the reasons to believe in the reality of entities which are not directly observable – which is the business of science – but on the normative evaluation of these reasons. I define epistemological scientific realism thus: we have better reasons to believe in the existence of some detected entities (sets of properties) and in the truth of some propositions about them rather than adopting an agnostic suspension of judgement. Detected properties are properties that are observable in principle, but which can only be observed with the aid of instruments.

Most scientific realists resort to abductive strategies to defend their position. According to the no miracle argument the success of science would be inexplicable, i.e. would be a miracle, if our best theories were not at least partially true about entities such as electrons, genes, etc. I argue that abductive explanationist strategies are unconvincing. As an alternative, I propose an inductivist strategy to defend a moderate version of scientific realism.

Do we have good reasons to believe that our best scientific theories are true? This old question, which hinges on the limits of scientific knowledge, is still passionately debated today. In this chapter and the next, I will argue in favour of a moderate version of epistemological scientific realism according to which we have more reasons to believe in the partial and approximate truth of theories that are empirically correct, especially in the truth of some scientific laws.

First, I discuss Bas van Fraassen's "loss-of-reality objection", which challenges the very possibility of our models to represent external entities, including directly perceived things. I show that there is no serious reason to doubt that some models

M. Ghins, *Scientific Realism and Laws of Nature: A Metaphysics of Causal Powers*, Synthese Library 483, https://doi.org/10.1007/978-3-031-54227-5_2

correctly capture certain aspects of external realities. On this I agree with van Fraassen. However, unlike him, I don't appeal to a pragmatic argument to *dissolve*, as he says, the loss-of-reality objection. Today, most philosophers look at science as being foremost a practice. Unlike them, I propose an approach which I call "theoretical". Within the perspective of this theoretical approach, I focus on *theories*. I will ask to what extent we have *more* reasons to believe in their (at least partial) truth rather than suspending our judgement about their truth or falsity. I will also emphasise that the main issue epistemological realism is confronted with is to assess the strength and weaknesses of realist and antirealist arguments. If we want to do philosophy, we must adopt a resolutely normative perspective, and not be satisfied with allegedly scientific[1] descriptions of scientific practice and theories. Thus, I oppose the version of naturalism according to which philosophy of science is nothing else than a science of science.

In order to defend the moderate and selective version of epistemological scientific realism that I favour, I do *not* implement an explanationist strategy as most scientific realists do. Explanationist strategies typically resort to some version of the celebrated "*no-miracle* argument" promoted by Hilary Putnam. In this chapter, I present a critique of this argument and the explanationist strategies usually deployed to defend scientific realism. Specifically, I show that even if the supposed existence of some entities that cannot be directly observed, such as atoms or viruses, provides the *best* available *explanation* of data, this does not give us better reasons to believe in their existence rather than to adopt a sceptical position or suspend our judgement about their existence.[2] As I will show in the next chapter, in order to be strong and convincing, the arguments in favour of the existence of entities such as atoms etc. must be similar to the arguments that justify our belief in the existence of directly observable entities. Such arguments, which are not explanatory, entitle us to believe in the existence of entities with properties that are indirectly observed and in the truth of propositions about them. The realism I support is a realism of instantiated properties in concrete things.

Before going any further, I must clarify what I mean by "observable property". In the class of observable properties, I include in the first place those properties that are directly observable or perceptible with our five senses, if not actually at least in principle. Some properties, although immediately observable, are not *de facto* directly accessible to us, either for practical reasons or because of the limitations of our senses. Dinosaurs and the planet Neptune have directly observable properties. If I lived in the time of the dinosaurs or if I were close enough to Neptune, I could directly observe some of their properties. I will also call "observable" properties

[1]This point is controversial. If science is an activity, it is obvious that science, like any other activity, can be qualified as good or bad, to the extent it achieves its objectives or not, or has desirable or unwelcome consequences. As an activity, science is not morally or axiologically neutral. I am inclined to believe that scientific practice is good to the extent that it aims at getting some kind of knowledge. But I will not discuss this issue here. On the other hand, theories, taken in themselves, can be considered to be axiologically neutral.

[2]For the sake of brevity, I will occasionally use "to know" instead of "to have better reasons to believe" when there is no risk of confusion.

which are too large or too small to be perceived by our unaided senses, for reasons that will be discussed later. For example, velocity is directly observable when it is neither too small or too large and instantiated in ordinary things. Similarly, a property such as position is directly observable for ordinary things, but not when instantiated in viruses, molecules or electrons. Yet I will take position and velocity to be observable properties in general. To call certain properties of viruses (such as position) "observable", in a broad sense, is, as I shall show, justified because they are epistemically accessible to us via causal chains inductively attested by means of instruments, such as a microscope, the reliability of which has been empirically and inductively confirmed. Observable properties which can be observed only indirectly, via causal chains, will be called "detectable". Detectable properties form a subset of observable properties.

Some properties lie beyond any possible observation, whatever new instruments could be invented in the future, simply because our perceptual apparatus does not change. These properties are observationally transcendent, so to speak. I will also call them "purely theoretical". Properties such as charm, strangeness etc. typically introduced in elementary particle physics are purely theoretical. From now on, when I mention unobservable properties, I will refer to purely theoretical properties only. In this and the next chapter, I try to defend a version of scientific realism about properties that are detectable – and detected – by means of causal chains the presence of which is empirically ascertained. However, we should adopt an agnostic position, that is, suspend our judgment with respect to the existence of purely theoretical properties.

According to the synthetic conception of theories, the issue of scientific realism can be addressed within two distinct perspectives, either model-theoretic or syntactic. The first relates to the arguments that give us good reason to believe that some *models* correctly represent their target, at least partially. The second has to do with accepting the truth of scientific *propositions*, in particular believing in the approximate truth of scientific *laws*. Within the first perspective, I will examine whether the modelling process makes us lose contact with real external targets. In my rebuttal of the loss of reality objection, I will show that our models do provide reliable information about external things since, as I have insisted in the previous chapter, their representational capacity is grounded on facts and true propositions that describe them. I will then be able to develop my argumentation in favour of epistemological scientific realism within the syntactic perspective, according to which theories are sets of propositions, as most authors do.

2.1 The Loss of Reality Objection

The issue of epistemological scientific realism arises at two distinct levels. At the first level, we must look at the relationship between models and perceived entities and ask whether we have reason to believe that directly observed entities such as flowers, clouds and stars – which are concrete things – can be represented by the structures we extract from them. At the second level, we evaluate whether we are entitled to believe that structures of properties that elude direct observation and are represented by mathematical models are, in fact, instantiated in real external entities. In recent decades, the debate on scientific realism has focused on the latter issue. Yet the first level, despite of being closest to experience and raising a fundamental philosophical issue, has rarely attracted the attention of philosophers of science. Bas van Fraassen deserves credit for addressing this unduly neglected question in the following terms:

> *How can an abstract entity, such as a mathematical structure, represent something that is not abstract,* something in nature? (2008, p. 240)

In my terminology, I will rephrase this question in the following way: how can a model, which is an abstract structure, represent a concrete real entity? This is a serious problem, which van Fraassen calls "the Loss of Reality Objection". If it remains unsolved antirealist positions such as constructivism, according to which we only know the models we construct, shall carry the day.

Before assessing van Fraassen's refutation of this objection, it must be acknowledged that constructivists are right when they contend that an abstract structure cannot, strictly speaking, represent a concrete entity. According to the structural conception of representation, a representational relation can only hold between *structures.* We succeed in representing only if we have constructed a representational function (an isomorphism or homomorphism) between the represented structure and the model it represents. Since a concrete entity is seen as a system, I cannot represent it as concrete by a structure. What then is the relationship between a *theory*, understood as a set of models, and the concrete perceived entities, the phenomena, that the theory is supposed to represent? Here is van Fraassen's answer:

> *For us* the claims
> A: that the theory is adequate to the phenomena and the claim
> B: that it is adequate to the phenomena as represented, i.e. *as represented by us,* are indeed the same! (van Fraassen 2008, p. 259)

To say that a theory is adequate to phenomena is to say that it contains empirical substructures that match the data models. In this case, the theory is empirically adequate to its observable domain (van Fraassen 2008, p. 250). Thus, *for us*, to represent phenomena is to represent them by means of empirical substructures isomorphic to data models obtained from phenomena. We cannot do otherwise. As a consequence, we cannot assert A and at the same time deny B, or vice versa. Such an impossibility does not arise from the contents of A and B: the concrete phenomena are distinct from the phenomena as represented by us. But we lack empirical epistemic access to facts which would possibly make A true and B false because, in practice, our action of representing phenomena is tantamount to representing a data model by an empirical (also theoretical) substructure. Thus, in practice, I cannot assert A and deny B. If I do, I become *pragmatically* incoherent.

Of course, other data obtained with more accurate instruments may force us to revise and correct our theory and its empirical substructures. We would then get *another* representation aimed at the same target. But these concerns are *epistemological,* whereas for van Fraassen the loss of reality objection is *ontological.*[3] If we claim that a theory is adequate to the phenomena as represented by our revised theory, we must also maintain that this theory is adequate to the phenomena, i. e. is an adequate representation of the phenomena. The crux of the matter however remains epistemological since it has to do with the cognitive import of the structures we construct. Do we lose epistemic contact with reality, that is, with a perceived concrete entity, when we represent it by an abstract empirical substructure?

For van Fraassen these concerns are groundless. Asking if a model is adequate to a phenomenal entity when we have constructed a representor of it would presuppose that we could bracket *our* doing the representation. A representation is always *my* representation, *my* representing action. We cannot just say that the model represents the phenomena. Representation is *not* a two-place relation, as we have seen in the previous chapter. My involvement must be taken into account. When representing, I am the one who constructs the representative function (at least implicitly) between a target seen as a system and the structure that represents it. I am never allowed to forget that I am the one who uses a structure to represent a perceived entity, always in some respects. A successful representation always involves an *indexical* element, an "I" that identifies its target and performs the representing action.

It is impossible for me to eliminate this indexical ingredient. A "god's point of view" or a "view from nowhere" from which I could independently contemplate the phenomena and my representational constructions is inaccessible. No *survol cartésien* is attainable by us. Being a consistent empiricist, van Fraassen then resorts to pragmatic considerations:

> *That A and B are the same for us is a practical tautology.* (...) this removes the basis for the Loss of Reality objection. (2008, p. 259)

For van Fraassen, in practice to affirm A is tantamount to asserting B. To deny this would put me in a situation of pragmatic incoherence. On this I quite agree with van Fraassen. Asserting proposition A while denying proposition B would be equivalent to saying that my theory is empirically adequate but that I don't believe that it is. Accordingly, van Fraassen offers not a solution but a *dissolution*, as he says, of the loss of reality objection. It is indeed impossible, in practice, to even formulate this objection because this could be done only from a godlike point of view.

It should be clear by now that the loss of reality objection does *not* hinge on a relation between structures. Our aporia has to do with the relationship between a model, such as an empirical substructure, and a phenomenon, such as an observed concrete entity. van Fraassen is of course well aware of this itch. However, I submit that the practical incoherence he detects does not allow us to escape from the closed world of our representations. According to me, if I represent a target as a system, *strictly speaking* I no longer represent the target, but the structure extracted from it.

[3] van Fraassen calls the loss of reality objection "metaphysical". However, the word "ontological" is more appropriate here since the present discussion unfolds at the level of representing phenomena and not unobservable entities.

Let me now reformulate the loss of reality objection using the terminology I introduced in the previous chapter. What I called the "perceptual structure" is obtained by abstraction-extraction from a concrete entity by adopting a certain point of view and focusing on selected properties and relations among them. It is indeed possible for me to extract *different* selected perceptual structures from the same perceived entity. All I have to do is vary the point of view I adopt. Subsequently, via measurement operations, I can construct various data models. For van Fraassen, "contents of observation or measurement outcomes" are "appearances" (van Fraassen 2008, p. 8). Thus, different representations can be extracted from the same observed concrete things (phenomena) according to the point of view of the user, an "I" in the first person who selects some properties and relations and omits others. In a way, phenomena are the potential source of various appearances. Phenomena do not contain a *unique* natural structure that is waiting in the wings to be represented by us. And on this point too I strongly agree with van Fraassen.

Let us have a look at a mundane example. A map of the Brussels subway network can be considered as a geometric structure. It becomes a concrete map or plan when it is appropriated by a user who associates some real properties, such as tracks and their intersections, with lines and points drawn on a piece of paper. The user also has to locate and orient himself with respect to the map.

We could then rephrase van Fraassen's affirmation above by applying it to a map:

> *For me* the claims
> A' that the map is adequate to the Brussels subway network.
> B' that it is adequate to the Brussels subway network as represented, i.e. *as represented by me,* are indeed the same!

I acknowledge, of course, that in practice I cannot both assert A' and deny B′, or vice versa, without being incoherent. I cannot believe in the truth of A' and, at the same time, deny the truth of B′. As we have seen, such practical impossibility results from my incapacity to positioning myself outside or above my own representational activity. In order to succeed in representing, I constructed a representational function, albeit implicitly, between selected properties of the subway network (the target) and some properties of the map (the representing artefact). In doing so, I also accepted the *truth of propositions that attribute these properties* to the subway network and the map, respectively. For example, the real intersection of two subway lines is made to correspond to a specific crossing of two printed lines on the map. If the propositions that describe those facts are true and an appropriate representative function has been constructed, the map is correct and contains useful information which, provided I correctly located and oriented myself on the map, helps me get to the intended destination. For example, if I am at the intersection of the blue and red subway lines, the map tells me how to reach the intersection of the blue and green subway lines, provided that the colours conventionally assigned to the subway tracks correspond to the colours of the lines drawn on the map.[4] Of course, we could have

[4]Evidently, subway tracks do not have blue, green etc. colours. The association of a colour with a subway line is a convention manifested by signs in the stations. "Blue" functions as an abbreviation

been interested in other aspects of the subway network such as the velocity of trains, the width of the tracks and so on.

Let's get back to gases. When a scientist says that a gas has a one litre volume, this is a claim about a perceived entity. The proposition <This gas is occupying a volume equal to 1 dm^3 > has the subject/predicate form <S is P>. The grammatical subject of the proposition is "this gas" and its predicative term is "occupying a volume equal to 1 dm^3 ". The predicate term refers to a property. When scientists state such propositions, they are *not performing* a representational action. Nor do they assert the correctness of a representation. When we attribute a property to an entity, we do not at all represent it as having a property, in the technical structural sense explicated in the previous chapter. In a predicative assertion I simply claim that a given thing instantiates a specific property, without claiming any kind of correspondence (let alone representational correspondence) between the thing and something else, be it an image in my mind, a thought or whatever. As a consequence, by making a predicative claim I do *not* create a gap between a representing artefact and a target, simply because *I do not represent.*[5]

Is it true that a perceived gas, for example the air in my office (the concrete target) has properties of volume, temperature and pressure? Yes. Is it true that these properties have a certain numerical value? Yes. Is it true that these properties have numerical values *a, b, c* in appropriate units? As far as I can tell, if I have carried out my measurements carefully, the answer is yes. Is it true that these numerical values hold in certain relations that approximately satisfy the Boyle-Charles formula? Yes. Even if the indexicality is always present, these positive answers can be given by several observers who perceive the same entity and make the same measurements. The indexicality of my representational activity does not prevent possible agreement between different observers or scientists, and thus objectivity in the weak sense of intersubjective agreement.

By answering "yes" to the four questions above, we are moving towards a realist position. If we pay due attention to the propositions whose truth grounds the success and the correctness of our representations, the loss of reality objection is not eliminated or dissolved as van Fraassen claims but is *met* within a realist perspective. *Reality is recovered for the simple reason that it has never been lost.* In fact, we do *not* represent things at all! Only structures can be represented. Things, on the contrary, are concrete entities with a wealth of properties. Thus, the concrete observed behaviour of a gas cannot be *stricto sensu* represented by perceptual structures, data models and empirical substructures. I of course concede that a gas is represented *as a* system in the sense that we have extracted a perceptual structure and a data model from it. But again, according to the structural conception of

for an expression such as "the track from station A to station B". A real property of a metro line is that it connects the stations at its ends as well as the intermediate stations.

[5] See Ghins (2010, p. 533).

representation, a concrete phenomenal entity cannot be represented, that is, represented by an extracted structure.[6]

Are we then condemned to fall back to a constructivist antirealist position, according to which we only know the structures we construct? Does scientific truth consist in the mere agreement between structures? Between a perceptual structure and a data model, for example, or between a data model and an empirical substructure? At the end of the day, this is where van Fraassen's position, which he appropriately calls "constructive empiricism" (1980), leads to. For him (but not for me), the connection between our representations and concrete reality is based on an indexical judgement, a first-person designation devoid of any cognitive content about the intended target. As van Fraassen puts it:

> in the chain [*theory*] – [*data model*] – [*reality*] the last link is one that is expressed in indexical judgements. (van Fraassen 2008, p. 257)

If this is so, van Fraassen's conception of representation is structural up to a certain point. It holds as far as my representation of a data model by an empirical substructure is concerned. But it stops at the level of concrete reality. There, the loss of reality is evaded by a pragmatic move. It does work. But the pragmatic obligation to assert A (or A') if B (or B′) is also asserted does not translate into a cognitive identity. As I said, I grant that I cannot assert that my representation is adequate to the target as represented by me without asserting at the same time that my representation is adequate to the target. However, since the two assertions do not have the same content (as van Fraassen acknowledges), they do not have the same cognitive import. My adequately or correctly representing the external target is not warranted by the theory correctly representing a data model, unless it is true that the target does have the properties organised in the data model. The indexical ingredient is certainly present (and on this I again agree with van Fraassen), since I am the one who selects the relevant properties and relations according to the point of view I adopt. But the fact that the target actually instantiates (or does not) the selected properties is totally independent of my doing the selection.

In my view, contact with external reality is achieved in direct perceptual experience. Immediate observation grounds the truth of propositions about the instantiations of properties in things. Despite its importance, I will not argue in favour of this point here. Let me just say that I am sympathetic with a form of direct realism with respect to immediately perceived properties, in the line defended by Brewer (2011).[7] We lose touch with reality when we erroneously believe that we are imprisoned in the cavern of our representations and the isomorphic functions we establish between them. Fortunately, we do not live in Plato's cave. If they are correct, our

[6] As we saw in the first chapter, my way of conceiving the relation of representation is not in line with the usual linguistic usage. It is common to say that a photograph represents a particular person or that a painting represents a landscape. Of course, we can continue to do this. I do not intend to regiment everyday language...

[7] Unlike Brewer who defends a direct realism of *objects*, I advocate a realism of (instantiated) *properties*.

representations are artefacts whose elements – such as numbers, functions or elements of a drawing or graph used in science – denote real properties instantiated in targeted observable entities and, on top of this, contains relations corresponding to relations instantiated in the target. Albeit abstract, models and structures are not mental entities, as van Fraassen rightly emphasises. They can be systems of sticks or mathematical signs. We certainly don't need to resort to a veracious Cartesian god to guarantee the correctness of putative clear and distinct mental representations to mind-independent realities. If we remain always aware that the success and correctness of our representations are based on facts and true propositions about them, we clearly (and also distinctly) realise that our contact with reality has never been lost.

The root of my disagreement with van Fraassen lies on diverging conceptions of truth. For van Fraassen, who agrees with Quine on this, the truth of the proposition <snow is white> rests on the reality of snow, and that's all (van Fraassen 2008, p. 248). According to Tarski's (1944) famous semantic definition of truth, the statement "snow is white" is true if and only if snow is white. Tarski's definition is hardly disputable. Yet it is compatible with a variety of philosophical conceptions of truth. According to Quine, claiming that a statement *is true* adds nothing to the mere enunciation of that statement. It's enough to just utter or write the statement. This is why Quine's conception of truth is also called "disquotational" or "redundant". It simply amounts to sticking to Tarski's definition, nothing more. It is very popular among strict empiricists. For my part, I will argue in the next section in favour of a correspondence view[8] of truth according to which the truth of the statement "snow is white" depends on the reality of an external fact, namely the instantiation of properties.

But van Fraassen explicitly rejects the correspondence view of truth (2008, p. 252) and seems to conflate truth with warranted assertability. If we are entitled to assert proposition A above, we are not allowed in practice to deny B, and vice versa. Even if A and B have distinct contents, we do not have cognitive access to hypothetical external facts that would make A false and B true. Granted, the notion of a reality *an sich* is unempirical and unintelligible and the representing activity is indexical. But, to repeat, we have direct access to phenomena, that is, observed properties instantiated in things. And we can make statements about them independently of our doing any representation. Moreover, such statements are objective and devoid of indexicality. If this is correct, B could be true and nevertheless A false, even though I would be pragmatically unable to say it. Indeed, a substructure of my theory could be adequate to the data model that I extracted from the phenomena, and yet the theory would not be adequate to the observed things, simply because the properties and relations of the data model are not actually instantiated in them.

But for van Fraassen we have no good empirical reasons to believe in the existence of real properties and relations. For him, such belief is meta-empirical

[8] I make a distinction between a correspondence *view* of truth and a correspondence *theory* of truth. The latter offers a description of the nature of the correspondence, whereas the former does not. More on this follows.

(metaphysical) and shouldn't be entertained. As I will argue in the next section, the truth of propositions depends on the instantiation of real properties and relations. Phenomena are perceived concrete co-instantiations of real properties. In the process of model construction, we organise some of their properties into structures. These real properties and structures make true the propositions on which our successful and correct representations are grounded. Contrary to what van Fraassen seems to affirm in some places (van Fraassen 2010, p. 548), the "substantive correspondence theory of truth", as he calls it, does not imply that nature as a whole has a single structure. I fully agree with him when he says that phenomena can give rise to a variety of representations that cut phenomena into various structures depending on the point of view adopted by a particular I. But this does not prevent some selected properties and some of their relations from actually being instantiated and thus making propositions about some facts true.[9]

2.2 The Realist View of Truth

My reply to the loss of reality objection would not be complete unless I make clear what I mean by the truth (and falsity) of a proposition. I submit that the *realist view of truth as correspondence* remains the most plausible. First of all, it is consistent with our use of the word "true" in most ordinary circumstances, and above all in science. The proposition <this gas has a one litre volume > is true (or false) in virtue of something which exists independently of my mind, my desires, my language, my beliefs, my arguments for believing or not believing in its truth, my personal history etc. A mental fact can undoubtedly be described by a proposition that is true or false according to what actually happens in my mind. I can truthfully describe a thought as pleasant, frightening etc. Nevertheless, the truth (or falsity) of a proposition about a mental fact is also "independent of my mind" in the technical sense that it doesn't depend on a wish, a desire etc. Granted, a mental fact is internal in the sense that it occurs "in my mind", but this doesn't prevent such fact to be *external to the proposition* that describes it, correctly or incorrectly.

According to what Charles Martin called the *truthmaker principle*, there are facts in the world which make true propositions true. Here is David Armstrong's formulation of this principle:

> (...) for every contingent truth at least (and perhaps for all truths contingent or necessary) there must be something in the world that makes it true (...) (Armstrong 1989, p. 88).
>
> The truthmaker is the 'correspondent' in the Correspondence theory of truth. (Armstrong 2000, p. 150)

To defend the truthmaker principle, Armstrong deploys an explanationist strategy. The occurrence of its truthmaker *explains* why a proposition is true. But the nature of

[9]See van Fraassen (2010, pp. 552–553) for his reply to these objections.

this explanation is far from being clear (see Daly 2005). Be it as it may, there is an alternative to explanatory reasoning here. The strongest argument in favour of the truthmaker principle is the experience of error. We frequently are forced to admit that we have to revise our beliefs. Ordinary experience often confronts us with phenomena that constrain us to readjust our beliefs, whether we like it or not. Our beliefs may *clash with reality*, as Karl Popper put it. Phenomena – perceived facts – can falsify propositions we believed to be true. Such perceptions impose themselves upon us, quite independently of explanatory reasoning. Thus, our experience of error[10] supports the falsemaker principle, and then the truthmaker principle, since a falsemaker makes true a negation of the proposition that we believed to be true. When a proposition faces the "tribunal of reality", it is true or false by virtue of an external reality, something "in the world", as Armstrong says, be it a truthmaker or a falsemaker.

Of course, in the experience of error my encounter with something unexpected explains (at least partially) why I gave up some of my beliefs. However, the case in favour of the truthmaker principle is not supported by explanatory reasoning but by experience. Such experience consists in perceiving something which we realise to be incompatible with previous beliefs. Such experience is a *meta*-experience since, in addition to being acquainted with phenomena, we become aware of a conflict between them and our beliefs. We are then led to accept that every true proposition is made true by something that "doesn't come from us", that is, an external truthmaker (or several of them for universal propositions). We thus end up with a realist conception or view of truth, according to which there is some sort of reciprocal correspondence between a true proposition and its real truthmaker.

So far so good. But the truthmaker principle is not immune from difficulties.[11] The main one consists in determining the nature of truthmakers. First of all, I will suppose, without giving an argument, that the truth of analytic propositions, including logical laws, rests on the meanings of their terms, such as "and", "or", "bachelors" etc. The truthmakers of these propositions are not in the world.[12] Furthermore,

[10] Alternative views of truth, such as the coherentist view, allow for the possibility of error of course. But these views must eventually rely to some form of correspondence. It would be too long to discuss this issue here. Let me just mention that the coherentist is confronted with incoherence only when she must abandon one of her previous beliefs (supposed to be coherent with the rest of them) by facing some facts. Then, she can adjust her system of beliefs in different ways to restore coherence. But the origin of such a "crisis" lies in some new fact which is independent and external to her.

[11] For an in-depth discussion of these difficulties, see Beebee and Dodd (2005).

[12] I will not broach on the possible existence of a truthmaker for mathematical propositions. I am mainly interested in the issue of scientific realism which hinges on the justification of belief in the truth of propositions which, even if their formulation is mathematical, do not have mathematical content only. I am exclusively concerned with synthetic propositions that relate to phenomena or facts that are empirically accessible in principle. Analytic propositions such as <All bachelors are unmarried> are true by virtue of the meanings of the terms they contain. Of course, the analytic/synthetic distinction is tricky, and we can be mistaken in attributing analyticity to a proposition. For example, most people will think that <only females can become pregnant> is an analytic proposition. However, this proposition is falsified by the way seahorses reproduce; it is thus synthetic.

I will argue in Chap. 4 that the truthmaker of a contingent (non-necessary) proposition such as <the rose is white>, is a fact, that is, the instantiation of a property F in a particular entity *a*, as Armstrong contends.

Negative propositions raise another difficulty for the truthmaker principle. Consider the proposition <the rose is white>. If the rose I am presented with is red, we have a falsifier of this proposition. This falsifier is also a truthmaker of the proposition <the rose is red>. The fact that the rose is red is also a truthmaker of the proposition <the rose is not white>. Obviously, a negative proposition has an unlimited number of truthmakers, but it is made true by a falsifier of the corresponding affirmative proposition.[13]

According to this realist conception of truth, when a proposition is true (or false), its truth is not subjective; it is objective since it depends on the existence of a fact "in the world". Objectivity does not boil down to several people intersubjectively agreeing on its truth. A proposition can be objectively true when everyone believes it to be false. There was a time when everyone believed that the earth did not move... Surely, beliefs are relative to the people who hold them. But, if the realist view of truth is correct, when a proposition is true it is *absolutely* true; and when a proposition is false, it is absolutely false. The truth of a proposition is *not relative* to my wishes, my beliefs, my language, my mental state, the reasons given for my belief in its truth etc. The truth of a proposition is indeed relative to some facts (its truthmakers...), but these facts exist independently of my beliefs etc. In this sense, the truth of a proposition is absolute.

Far from committing us to dogmatism, the realist conception of truth is the most effective antidote to it. Dogmatists insist that some of their beliefs are beyond possible revision. For them, there cannot be any reason to question their beliefs or to change their mind in any circumstance. New factual information or arguments will leave them cold. In fact, dogmatists do not possess "absolute truth". They just claim to be *absolutely certain* that some propositions are true. Certainty is a mental, psychological state which is relative to individual persons. But such psychological certainty does not provide any justification for belief in truth. Surely, I am certain that I am now sitting at my desk. But this belief is justified by perceptual experience that gives me reason to have such belief. Moreover, such beliefs are vulnerable to traditional sceptical arguments (I could be dreaming, for example). As for general propositions, which are of utmost importance in science, new facts can always occur which would compel to revise my belief in these propositions. On the other hand, if truth is absolute, this provides an incentive to try to discover new facts, which may confirm, but also falsify, our beliefs. Dogmatists on the contrary erroneously believe

---

[13] The truthmaker principle seems to be challenged by true propositions such as <there are no penguins at the North Pole> (Beebee and Dodd 2005, p. 3). But if we analyse such a proposition *à la Russell*, we get: <it is false that there is something that has the property of being a penguin and the (relational) property of being at the North Pole>. All penguins are at the South Pole and anything at the South Pole cannot be at the North Pole at the same time. This fact falsifies the proposition <there are penguins at the North Pole> and makes its negation true.

that it is pointless to confront some of their beliefs with facts. They thus rest satisfied with epistemic immobility.

Actually, dogmatism eventually leads to scepticism. According to dogmatism, genuine knowledge must be certain and indubitable. As a consequence, given that certainty cannot be justified, knowledge cannot be reached. A conception of knowledge that requires justified certainty entails scepticism. Such kind of knowledge is inaccessible because absolute certainty is beyond our reach. Surely, knowledge is difficult to attain. But genuine knowledge must be true. False knowledge is an oxymoron, a contradiction. Granted, we can never be "absolutely" sure that we know some facts. However, we have sometimes better reasons to believe rather than not in the truth of some propositions on the basis of direct perception in normal circumstances or sound[14] arguments. According to the realist conception of truth, truth cannot be conflated with justification, however good it may be. A justified proposition can always be falsified by external facts. Truth transcends justification. A truthmaker, a fact, is independent of the reasons we might have to believe in its existence.[15] Then, the main epistemological problem hinges on the justification of our belief in the truth of a proposition. What counts as a *good* justification? This vexing but crucial issue will be dealt with at length in the next chapter.

Let us return to the correspondence *view* of truth. For the sake of completeness, aren't we required to spell out the nature of correspondence between a true proposition and its truthmaker? In other words, are we required to elaborate a *theory* that would tell us what this correspondence consists in? Actually, a correspondence *view* of truth remains defensible even if we are in the dark about the nature of such correspondence. This is why some authors like Evandro Agazzi (2014) prefer to speak of a "referential" conception of truth, according to which a true proposition *refers* to its truthmaker. Such conception is also defended by Gottlob Frege.

A correspondence *theory* of truth, on the other hand, should provide an explication the correspondence between a true proposition and its truthmaker. Such a theory was developed by Wittgenstein in his *Tractatus Logico-Philosophicus* first published in 1921. For Wittgenstein, a true proposition[16] is a "logical image of facts" (1971, p. 19). The correspondence here consists in an isomorphism between a proposition and a possible situation. If this possible situation is real, the proposition is true. A true proposition then represents a fact, in accordance with the structural conception of representation developed in the previous chapter. When it represents a real situation – a fact – the proposition is true. The well-known failure of this attempt – a failure that Wittgenstein himself quite elegantly acknowledged – shows that propositions do not represent, in the technical (structural) sense in

[14]Recall that an argument is logically *valid* when its conclusion is necessarily true *if* its premises are true. A *sound* argument is a valid argument whose premises are all true.

[15]On the distinction between truth and justification, I recommend Diego Marconi's excellent book (2007).

[16]In fact, Wittgenstein says that a *thought* is a logical image of facts. Since he also says that what makes a thought accessible to our senses is a statement or sentence (*Satz*), I interpret thought as what I mean by proposition. (I do not follow Pears and McGuinnes in translating *Satz* by "proposition").

which I understand the term "representation". That is why I do not think that our (true) propositions represent reality. Language does not represent the world. To my knowledge, we do not (yet?) have a satisfactory *theory* of truth as correspondence. But this does not undermine the realist *conception or view* of truth, based, as we saw, on our experience of error.

2.3 Epistemological Scientific Realism: The Main Issue

Bas van Fraassen wisely reminds us that the philosopher's task is not to justify belief in the existence of facts.

> Do electrons exist? Are atoms real? These are not philosophical questions.
>
> Whether electrons exist is no more a philosophical question than whether Norwegians exist, or witches, or immaterial intelligences. Questions of existence are questions about matters of brute fact, if any are, and philosophy is no arbiter of fact. (van Fraassen 2017, p. 95)

It is up to scientists, not philosophers, to argue for the existence of electrons and atoms. In the debate on scientific realism, philosophers are concerned with the *normative task of evaluating the strength of the kinds of arguments* that are claimed to *justify* belief in the reality of certain entities and the truth of statements about them. For example, it is for the philosopher to assess the legitimacy of a form of inference known as inference to the best explanation, to which most scientific realists attributes the capacity to justify belief in the existence of some indirectly observable entities. Examining the grounds of the strength (or weakness) of *kinds* of arguments is an epistemological – and also logical – issue. It deserves to be thoroughly examined because it hinges on the level of confidence we are entitled to place in our best scientific theories, in general. Even if the arguments in favour of belief in the existence of electrons are obviously different from those invoked to justify belief in the reality of mitochondria, these arguments share common features that make them sound or solid and therefore convincing. This is what the scientific realist upholds. Philosophers are not especially interested in specific arguments, but in the grounds for the soundness of kinds of arguments. Scientific realists then face the philosophical challenge of identifying the *grounds* of the cogency of the kinds of arguments offered in favour of belief in the existence of entities inaccessible to direct observation.

This investigation will certainly shed light on *what science is*. Science is supposed to be knowledge. How far does it reach? What are its limits? Should justified belief in the truth of scientific theories be limited to what these theories say about what is – directly, immediately – observable by our five senses, or can it be legitimately extended, as I shall argue, to entities whose properties, although observable in principle, are only *in*directly accessible to us, i.e. detectable? There can be little doubt that determining the limits of justified belief in the truth of scientific theories is a genuinely philosophical task, which echoes obvious Kantian resonances.

To this central philosophical question – *what is science?* – van Fraassen gives a resolutely *pragmatic* answer. His first task then is to specify the goal of scientific activity. By doing so, the fundamental criterion of success for scientific practice is identified (van Fraassen 2017, p. 98). A practice is good if it leads to success, that is, the achievement of some objective. Success is evaluated by the degree of achievement of the goal pursued. If I decide to play the violin, my goal will be to play beautiful music. The success of this activity will be judged by the quality of the music produced. In the same way, the success of scientific practice will be assessed by comparing the results obtained with the objectives pursued.

Thus, a philosophical answer to the question "what is science?" must first specify what the aim of scientific activity is. For van Fraassen, scientific realism is philosophical attitude or stance according to which scientific practice aims at the construction of *true* theories. In line with his pragmatism, van Fraassen adds that acceptance of a theory by a scientist involves commitment to *work* within the framework provided by the theory. But acceptance involves some beliefs. According to scientific realism, these beliefs are not limited to immediately observable phenomena, but can legitimately be extended to include beliefs in entities inaccessible to direct observation.

> Science aims to give us, in its theories, a literally true story of what the world is like; and acceptance of a scientific theory involves the belief that it is true. (van Fraassen 1980, p. 8).

This famous quote from *The Scientific Image* is followed by the terse sentence: "This is the *correct statement* [emphasis is mine] of scientific realism."

Since van Fraassen's characterisation of scientific realism is presented as a correct statement, not a definition, what would make it true (or perhaps false)? Science is practiced by human agents: scientists. Do they seek to discover truths about the world? When they accept a theory, do they *ipso facto* commit themselves to belief in its truth? To answer these questions, the only way to proceed is to make a survey and ask scientists what their goals and beliefs are when they engage in scientific activity. If most participants in the survey answer "yes" to both questions above, then we can conclude that scientific practice aims in fact at producing true theories and that scientists, at least most of them, do believe in the truth of the theories they accept (which does not, of course, imply that these theories are true...). On the other hand, if most participants in the survey declare that they just try to save phenomena and that, when accepting a theory, they only believe in the truth of the observational propositions it contains, then the statement of scientific realism proposed by van Fraassen is falsified.

This is certainly *not* what van Fraassen has in mind. An argument for (or against) a philosophical position cannot be based on facts, such as the goals actually pursued by scientists and the beliefs they entertain. For van Fraassen, scientific realism is a *practical stance* (van Fraassen 2002) which is a possible answer to the central philosophical question "what is science?". And such answer is not a factual statement which could be true or false. Nor does van Fraassen understand his statement of scientific realism as a definition, like the definition of a circle or a bachelor. Scientific realism is a specific attitude towards scientific practice which is characterised by specific goals, beliefs and commitments. According to scientific realism, success (or failure) of scientific endeavour must be evaluated in relation to the aim pursued, namely the production of true theories. Good scientific practice aims at truth. Scientific realism appears to be a normative, even ethical, posture. For the realist, a scientist is a good practitioner of his trade, if she strives to construct true theories and accepts only the theories she believes to be true.

What van Fraassen means by the word "correct" in the present context is made clearer if we look at the statement of the opposite – antirealist – position he defends:

> *Science aims to give us theories which are empirically adequate; and acceptance of a theory involves as belief only that it is empirically adequate.* This is the statement of the antirealist position I advocate: I shall call it *constructive empiricism.* (van Fraassen 1980, p. 12)

As we just saw, scientific activity must be characterised in terms of goals, success, commitments and beliefs. And van Fraassen recommends a way of doing science which he deems better than what the realist proposes. He thus advocates an ethics of science, even if he doesn't say it explicitly. Moreover, van Fraassen is careful not to fall into the trap of caricaturing his opponent's attitude in order to win an easy victory. To that effect, he proposes what he presents as a "minimal" statement of scientific realism, which is weaker, and thus harder to refute, than more restrictive statements. Nevertheless, Fraassen is keen to preserve the core of scientific realism, which is the concern with truth. "Correct" then here means "minimal". More restrictive formulations are incorrect because they are more vulnerable to antirealist objections.

Why does van Fraassen recommend constructive empiricism instead of scientific realism as the best stance or attitude? Why is it better for a scientist to behave according to the norms prescribed by constructive empiricism? To defend his normative position, van Fraassen deploys a two-step strategy. First, he argues that the explanationist arguments based on inference to the best explanation, which most scientific realists deem to be cogent, do not justify belief in entities that cannot be directly observed. Therefore, one cannot demand that accepting a theory implies belief in its whole literal truth, when there is no good reason to believe in the truth of propositions about things that cannot be directly observed. If this is so, scientists are unable to justify belief in the truth of a theory. The realist's criterion of success is inapplicable in practice since it is impossible to decide whether truth about entities that cannot be directly observed has been reached.

Secondly, van Fraassen points out that it is less risky, and therefore *pragmatically more rational*, to embrace constructive empiricism and commit oneself to believe in the truth of propositions about phenomena only. Indeed, believing only in the empirical adequacy of a theory is a more prudent attitude than believing also in what the theory says about things that elude immediate observation. Surely, it is not irrational to adopt scientific realism since "(...) what it is rational to believe includes anything that one is not rationally compelled to disbelieve." (van Fraassen 1989, pp. 171–172) For the constructive empiricist however, the only epistemic commitment involved in accepting a theory is the belief that it is empirically adequate. Given that such epistemic commitment is more limited than the one of the scientific realist, constructive empiricism can be considered to be a pragmatically more rational stance than scientific realism. According to van Fraassen, science is a practice which aims at saving phenomena only. Success of such activity is decidable since saving phenomena is an empirically verifiable goal.

No one would deny that science is also an activity. But the alternative – theoretical – approach I propose is to look at the theories produced by scientific activity and to assess the strength (or weakness) of the forms of argumentation adduced in favour of the existence of some entities that are not directly observable. This way of approaching the central question "what is science?" is synchronic, rather than diachronic since it focuses on outcomes of scientific activity and not on a practice that unfolds in the course of time. My approach can also be described as photographic, rather than cinematographic.

As a consequence, the main philosophical question that I will address is the following: is it *more* rational to believe in the truth of some scientific propositions about indirectly observed entities than to suspend our judgement as to the truth of these propositions? Epistemological scientific realists give a positive answer to this question. For them, it is possible to invoke *better* reasons for believing, rather than not, in the truth of *selected parts* of empirically correct theories, including selected parts of theories accepted in the past, even though these theories have been falsified and rejected by scientists today.[17] Scientific realists must meet the challenge of specifying the kind of arguments that legitimately justify belief in the truth of propositions about entities detected by using reliable instruments, despite their inaccessibility to direct observation.

2.4 Epistemological Scientific Realism: Definition

Assessing the strengths and weaknesses of the arguments for belief in the existence of some entities and the (approximate) truth of propositions about them is undoubtedly an epistemological issue. My approach will be *theoretical*, as I said, since it focuses on specific products of scientific activity, namely theories. Accordingly, I will direct my attention to the *reasons* to believe in the truth of *certain parts of* these theories while setting aside the objectives pursued as well as the commitments involved in scientific activity. These reasons will be assessed within the framework of this theoretical approach, which is resolutely normative, thus genuinely philosophical. Although I distance myself from van Fraassen's pragmatist perspective, I share his rejection of naturalized epistemology.[18] Philosophy of science should not be reduced to a science of scientific practice that would be restricted to the

[17] See Psillos (1999). Among falsified theories, it is useful to distinguish theories which are partially or approximately true such as classical mechanics and geometric optics (which are still taught today, and rightly so) and theories that have been completely rejected, such as Ptolemaic astronomy.

[18] On this, see van Fraassen (1995).

description of the behaviour of scientists and the scientific arguments they use to justify their claims. According to this view, philosophy of science claims would be true or false to the extent that they conform or not to actual scientific practice. Philosophy of science would then be reduced to a sociology of science perhaps enriched with a psychology of scientists. To be worthy of its name, philosophy is expected to make a case in favour of norms, as I insisted, which in the present case would allow us to evaluate the defects and merits of kinds of argumentations.

For the realist, the challenge consists in identifying the kind of arguments which give us more reasons to believe in some detected facts (i.e. instantiated properties), which are related to directly observed facts by empirically verified causal relations. I formulate the version of scientific epistemological realism (SER) I favour thus:

> It is more rational to believe in the existence of scientific properties which are not immediately observable but have been detected, rather than to suspend our judgement as to their existence or to assert that they do not exist.
>
> It is also more reasonable to believe in the (at least partial and approximate) truth of propositions (such as scientific laws) that describe relationships between these detected properties, rather than not to pronounce on their truth or falsity, or to assert that they are false.

According to this formulation, scientific realism is an epistemological, and undeniably philosophical, thesis. The debate about scientific realism hinges on the justification of beliefs (while admitting the possibility of error) in selected scientific propositions that attribute properties to entities inaccessible to direct observation. In what follows, I will support an *ontological* realism about the existence of detected properties whose instantiations make some propositions – scientific laws among others – true.

The above formulation of epistemological scientific realism consists in affirmations that are true or false. It could happen that no argument for believing in the existence of detected properties is strong enough, as claimed by strict empiricists like Bas van Fraassen. Nevertheless, I will attempt to show that arguments of a specific kind are compelling. For realists, the criterion for the success of scientific activity is the construction of theories that we have excellent reasons to believe to be true regarding detected entities. Only an epistemological argumentation can normatively justify belief in the success of scientific practice in this sense. This argumentation will be further developed in the next chapter.

As far as *certainty* is concerned, neither constructive empiricism nor scientific realism provide good reasons to be sure that a theory is empirically adequate or partially true. Nobody can justifiably be certain that a theory is empirically adequate, i.e. true about the directly observable phenomena that fall within its domain, simply because novel (perhaps unexpected) phenomena that fail to be embeddable in the theory might be discovered in the future. Scientists can never be absolutely certain to have reached their objective according to the criterion of success heralded by constructive empiricism. However, scientists (like everyone else, for that matter...) do have better reasons to believe in the empirical adequacy of a theory whose predictions are correct *so far*, rather than to suspend their judgement about its future adequacy, at least if some form of inductive reasoning is deemed reliable (as I will contend in the next chapter).

To avoid confusion, I will use the terms "predictive truth" or "predictive correctness" instead of the commonly used "predictive success". I make this choice because it seems more appropriate to attribute success to an action rather than a theory, since a theory is not an action.[19] (Avoiding explicit reference to success when dealing with theories is not an innocent change in terminology. It indicates a reversal of perspective: the passage from a pragmatic approach to a theoretical approach, as I emphasised earlier.) Now, observational truth (or observational correctness) appears to be akin to what van Fraassen calls "empirical adequacy". However, observational correctness is more restrictive. While an empirically adequate theory is expected to save *all* phenomena within its domain, including those not yet discovered, empirical correctness only pertains to currently available observations.

For scientific realists, if there are better reasons to believe in the truth of propositions about detected entities, there are *ipso facto* better reasons to believe in the truth of all their empirical consequences, including propositions that describe future observations. Granted, scientific realism is more risky than constructive empiricism since the former claims that justified beliefs are not circumscribed to directly observable entities but can be extended to detectable entities. As far as belief in entities that are not directly observable is concerned, error can result from a misevaluation of the strength of the reasons for believing in the instantiation of detectable properties, that is, an overestimation of the reliability of a justification procedure. This kind of error would be a consequence of a flaw in a philosophical analysis, which is independent of facts.

For sure, propositions about detectable entities used to derive predictions shown to be incorrect can no longer be considered to be true. Nevertheless, it is clear that the occurrence of falsifying observations or measurements is a matter of fact. As such, it doesn't threaten the cogency of an epistemological argument in favour of belief in detectable entities mentioned in propositions supposed to yield correct empirical predictions. For example, when the soundness of arguments in favour of the existence of electrons or the planet Neptune is discussed, the debate is conducted within the framework of theories that are supposed to be empirically correct. If such empirical correctness is undermined by observations, any realist argumentation becomes pointless. Empirical correctness is a necessary condition, albeit not sufficient, for justified belief in propositions about detectable entities.

In the next chapter I will try to show that the arguments in favour of belief in the existence of some detected entities, such as electrons or the planet Neptune, are similar to those adduced in favour of belief in the reality of entities such as pineapples and mice, which can be directly observed, but are not in some circumstances. These two categories of arguments have the same strengths, but also the same weaknesses, and do not eliminate the possibility of error. However, in the debate on scientific realism, antirealists and realists alike do not question the

[19] Recall that in ancient Greek θεωρια means vision or contemplation.

existence of ordinary, directly perceived, entities such as pineapples and mice, although it is obvious that our beliefs about such entities can be mistaken.

Usually, the construction of a theory that yields accurate predictions – including new and unexpected ones – and also manages to explain them is rightly considered a successful achievement. There is no doubt that many scientists strive to construct theories and models that make correct predictions and provide explanations, typically of a causal nature. This is a fact I will take for granted. When it comes to explanation, I will use the term "explanatory power" instead of "explanatory success", which aligns with my theoretical approach and the way some philosophers of science, such as Stathis Psillos, characterize scientific realism:

> The thrust of this realist [epistemological] thesis is that science can and does attain theoretical truth no less that it can and does attain observational truth, where by "theoretical truth" we understand the truth of what scientific theories say about unobservable entities and processes (. . .) (Psillos 1999, p. xx)

In accordance with my formulation of scientific realism, I will attempt to show in the next chapter that we have good reasons to believe that scientific theories do attain theoretical truths provided these describe detected entities. But before doing that, I will present a critique of the explanatory arguments in favour of scientific realism.

2.5 The No Miracle Argument

Most scientific realists rely on the "no miracle argument" which was made famous by Hilary Putnam.[20] Additionally, most scientific realists adopt naturalism and consider realism to be a *scientific* hypothesis. They argue that realism is the only hypothesis capable of *explaining* an observed fact, namely, that scientists consistently produce predictively correct theories. In Putnam's words, scientific realism provides the best scientific explanation for the success of science and is, in fact, the only acceptable one. The sole possible alternative would be the sheer absence of a scientific explanation. If scientific theories are not assumed to be at least partially true about some entities that cannot be directly observed, their predictive correctness would appear as a type of miracle or, as Jack Smart calls it, the result of an extremely unlikely "cosmic coincidence" (Smart 1963, p. 39). Miracles certainly do not qualify as acceptable scientific explanations; therefore scientific realism is true.

As Putnam says:

> (. . .) realists typically argue *against* some version of idealism – in our time, this would be positivism or operationalism. (. . .) And the typical realist argument against idealism is that it makes the success of science a *miracle* (. . .) The modern positivist has to leave it without explanation (the realist charges) that 'electron calculi' and 'space-time calculi' and 'DNA calculi' correctly predict observable phenomena if, in reality, there are no electrons, no

[20] Before Putnam, a similar argument was used by Grover Maxwell (1962) and Jack Smart (1963).

> curved space-time, and no DNA molecules. If there are such things, then a natural explanation of the success of theories is that they are *partially true* accounts of how they behave.
>
> (. . .) That science succeeds in making many true predictions, devising better ways of controlling nature etc., is an undoubted empirical fact. If realism is an *explanation* of this fact, realism must itself be an over-arching scientific *hypothesis*. And realists have often embraced that idea, and proclaimed that realism *is* an empirical hypothesis. (Putnam 1978, p. 18).

Putnam claims that the existence of some entities that cannot be directly observed and the truth of assertions about them is the only reasonable explanation for the success of scientific theories. In what follows, I will understand success in terms of predictive correctness while leaving aside pragmatic successes of science such as "controlling nature" etc. The no miracle argument is even praised by Putnam as the "ultimate", knockout argument against antirealists!

I will attempt to show below that no version of the no miracle argument is able to support epistemological scientific realism. But before doing that, we must be clear as how the no miracle argument works. To begin, it must be pointed out that it is a *top-down* argument which compares two competing explanatory hypotheses – the one appealing to truth, the other one to miracles – to conclude that since the first one is the only plausible one it is also the one which is true. The no miracle argument is a particular case of inference to the best explanation or abduction. On the contrary, a *bottom-up* argument attempts to show that a correct hypothesis is inductively supported by the facts.[21] Second, notice that this argument is a *meta-argument*. A good scientific theory predicts observations or measurements by means of hypotheses about unobserved entities such as electrons, genes, tectonic plates, and so on. The facts correctly predicted by a theory are *first-order level* facts F. The fact that a particular theory makes correct predictions is a *second-order level* fact F*. Proponents of the no miracle argument argue that in order to explain this second-order level fact F*, we must admit that some theoretical assumptions (about some entities that cannot be directly observed) are true.

Most people rightly consider the no miracle argument as a particular case of inference to the best explanation (IBE), which I will also call "abduction" in the context of the present discussion. Abduction goes as follows:

1. F is a fact
2. Hypothesis H[22] explains F
3. No other available hypothesis explains F as well as H does
   Conclusion: H is true.

Arguments of this form are invalid because in the first place it can't be excluded that an alternative hypothesis, incompatible with H – but unknown to date – explains fact F better. However, even if it can be proved that there is only one acceptable explanation of some fact (as Putnam claims) such arguments remain invalid because it does not contain a premise asserting a connection between (best) explanation and

---

[21] In Chap. 3, I will offer an inductive bottom-up defence of scientific realism.

[22] Abductive reasoning can obviously be also applied to a set of hypotheses.

truth. Is it possible to repair the no miracle argument and make it valid as well as sound, by adding a premise – assumed to be true – in the same way that some have tried to make induction logically valid by adding the assumption that nature behaves uniformly?[23] This is exactly what Alan Musgrave proposes to do by adding premise 1* to the abductive argument form above:

> 1*. "*It is reasonable to believe* that the best explanation of any fact is true
>
> 1. F is a fact
> 2. Hypothesis H explains F
> 3. No available competing hypothesis explains F as well as H does
>
> Therefore, *it is reasonable to believe* that H is true." (Musgrave 2017, p. 80)

Adding premise 1* to the no miracle argument, we get:

> 1*. It is reasonable to believe that the best explanation of any fact is true
>
> 1. Meta-fact F*: H is predictively correct
> 2. Hypothesis H* (H is true) explains F*.
> 3. No available competing hypothesis explains F* as well as H* does.
>
> Therefore, *it is reasonable to believe* H*, that is, to believe that H is true.

Before examining the soundness of Musgrave's reformulation, let me first observe that it is not the mere rationality of a belief that is at stake here. The realist must prove more. van Fraassen even asks the realist to show that "it is unreasonable *not* to believe that H is true." (van Fraassen 2017, p. 102) By requiring this van Fraassen poses a formidable challenge to the realist. In fact, it is impossible to provide arguments that would *force* someone to believe, on pain of irrationality, in the existence of entities that cannot be directly observed, just because the possibility of error can never be completely eliminated. On the other hand, it is possible to defend, as I will try to show in the next chapter, the *weaker* claim that it is *more* reasonable to believe in the truth of theoretical assertions when they speak of entities whose properties are detectable – and detected.

Returning to Musgrave's reformulation, we must readily acknowledge that the additional premise he proposes makes the no miracle argument deductively valid. But does it make it sound? Specifically, do we have good reasons to believe that premise 1* is true? As I will show, we do not.

To start, let me introduce an important distinction between two kinds of explanations, which is analogous to the distinction between valid and sound arguments. A sound argument is a valid argument all the premises of which are true. By analogy, a *correct* or *true* explanation is a *satisfactory* explanation whose all premises are true. An explanation is satisfactory if it gives us a good *understanding* or comprehension of the phenomena, but it might be false. For example, the stability of the motions of

---

[23] To put it briefly, a simple inductive reasoning goes the following way. Suppose we have observed so far that a finite set of things x with property F also have property G. We thus have a finite set of true singular propositions about particular things. By induction we infer from these premises the truth of the universal proposition: <*all* things in the world with property F also have property G>.

the planets is satisfactorily explained by the regular rotation of crystalline spheres, as proposed by Ptolemy. We clearly understand that if planets are attached to hard spheres in stable and regular motions, their trajectories in relation to the stars will remain unchanged over time. This is what we observe, at least for relatively short periods of time. However, Ptolemy's hypothesis is false and so is his explanation.

A satisfactory explanation must conform to rational criteria, for example it must be free of contradictions. But this is not enough. Additional criteria must be added. These criteria vary according to the different views of explanation that could be adopted. If an explanation is more satisfactory it has more explanatory power, but it could be false. If we fail to distinguish between satisfactory and true explanations, we expose ourselves to confusion. Proponents of inference to the best explanation do accept this distinction, since they admit that there may be several competing satisfactory explanations. Within a top-down approach they consider the internal merits of various available explanations to finally conclude that the explanation that is most satisfactory among them, namely, the one with the greatest explanatory power, is true or at least more likely to be true.

At this point, we are confronted with the classical problem of theory choice. How do we choose between competing theories or hypotheses? Pragmatists, such as Bas van Fraassen, recommend working within the framework of the theory that has more of the virtues that are deemed to be important. Explanatory power is one of them. Scientists should then commit themselves to working with the hypotheses that provide the best explanation of the data, while refraining from believing in the truth of these hypotheses. Indeed, even if we all agreed on the criteria used to compare the explanatory powers of competing hypotheses and all the possible explanations of some data were available – which is of course totally unrealistic – we would still have *no* good reasons to believe that the most satisfactory explanation is true. Why? Simply because *the explanatory power of a hypothesis does not support belief in its truth.*

This is the main difficulty of any explanationist approach. It is certainly not irrational to believe in the truth of a set of hypotheses that correctly predict phenomena, as long as they do not contradict hypotheses already held to be true. But is it *more* rational to believe that nature is organised in conformity with what we consider to be a good explanation, according to our own standards of understanding and intelligibility, and this even assuming that these standards are universally shared or built in our human nature? What connection to reality could be warranted by the explanatory power of a hypothesis, given that it is evaluated on the basis of our subjective[24] requirements of comprehension? In my opinion, none. As Peter

[24] Remember Peirce's famous formulation of abduction in his 1903 Harvard lectures:

> The surprising fact, *C*, is observed.
> But if *A* were true, *C* would be a matter of course.
> Hence, there is reason to suspect that *A* is true. (Peirce 1931, vol. 5, p. 189)

The expression "a matter of course" clearly refers to a subjective judgement, be it individual or collective. Abductive reasoning for Peirce had mainly a pragmatic, heuristic role of selecting hypotheses worth investigating.

Lipton asks: is the *loveliest* explanation, the one that pleases us most (and in this sense is the most satisfactory for us), also the one that is the *likeliest*, that is, the one that has the best chance of being true? (Lipton 2004, p. 61) Is there a pre-established harmony between our explanatory requirements and the organisation of nature? Such a Leibnizian harmony between nature and the explanatory requisites of our minds might very well obtain (who knows?), but how can we convincingly argue for it? This is what Lipton refers to as "Voltaire's objection":

> (...) supposing that loveliness [of an explanation] is as objective as inference (...) What reason is there to believe that the explanation that would be the loveliest, *if it were true* [underlining is mine], is also the explanation that is most likely to be true? Why should we believe that we inhabit the loveliest of all possible worlds? (Lipton 2004, p. 70)

Unlike Lipton, I do not think that an explanation would be the loveliest *if* it were true or correct. An explanation can be the loveliest or the most satisfactory without its premises being true, and vice versa. I want to keep loveliness entirely separated from truth. But I agree with Lipton on the main point: there is no relation, and much less harmony, between the loveliness or beauty of an explanation and real facts in the world. To believe this would be to fall again into the trap of some idealist prejudice[25] according to which our requirements of understanding give us cognitive access to real entities. Bas van Fraassen is quite right in emphasising that there is no connection whatsoever between explanatory power and truth. If this is correct, the no miracle argument cannot be rescued even if we accept that scientific realism is the *only* explanation of the predictive correctness of scientific theories. The epistemic divide between explanation and truth makes such rescue impossible. Abductive arguments may well play a *heuristic* role in devising new possible explanatory hypotheses, as Peirce stressed, but their explanatory force cannot provide a justification for belief in their truth.

Another attempt to rescue abduction in general, and then the no miracle argument, would be to point out that this form of reasoning has often been successful in the past. Such move would provide abduction some support, which of course would be inductive. Two remarks are in order here. First, as I will argue below, the apparent force of some abductive reasonings rests on previously inductively confirmed assumptions. Second, alternative plausible explanations do not come "out of the blue". They are not allowed to contradict well established theories and beliefs that have been widely confirmed inductively. Thus, possible historical inductive support of abductive reasoning would not vindicate the alleged truth-conducive power of loveliness and abductive reasoning as such, but would deliver, I suspect, more evidence in favour of induction as a reliable argumentative tool for justifying beliefs.

---

[25] Ironically, although Putnam uses the no miracle argument to counter idealism, his argument has an idealist flavour. Of course, the idealists targeted by Putnam are the empiricists who contend that only our perceptions or *sense data*, i.e. mental entities, are real. (van Fraassen is not one of these empiricists, since for him sense data simply do not exist (1980 p. 72) whereas ordinary perceived entities do exist.)

Could it be that the no miracle argument (without Musgrave's additional premise), although being a meta-argument, is *scientific* after all? If so, it would enjoy the same prestige and credibility as the arguments used in science. The no miracle argument clearly has the same logical form as that of an inference to the best explanation. And such form of inference is widely used in science. Which is hardly surprising since it has led to correct – sometimes spectacular – predictions but also, as it was to be expected, to false ones. Michael Levin (1984) and I (Ghins 2002) have argued that truth cannot play an explanatory role analogous to that of electrons or DNA. Truth is certainly not an entity, nor is it an explanatory ingredient which could reasonably be taken to be scientific. Therefore, scientific realism cannot in any way be equated with a scientific hypothesis. Yet this is what Putnam does when he says that "realism must itself be an over-arching scientific *hypothesis.*" (Putnam 1978, p. 18) In my view,[26] scientific realism is a genuinely philosophical – epistemological – thesis that does not aim to explain *facts*, whether they are first or second-order level facts.

Naturalist realists, such as Richard Boyd (1981), Stathis Psillos (1999) and Howard Sankey (2008), oppose this view. They have developed a science-based argument for the reliability of inference to the best explanation on the basis of the fact, acknowledged by all, that predictively correct scientific theories are available but also on the problematic assumption that particular theories are approximately true. As Psillos says:

> The (first-order) instances of explanatory reasoning involve the claim that it is reasonable to accept that *particular* theories are relevantly approximately true.[27] The no miracle argument is, then, based on these instances to defend the more general claim that science *can* deliver *theoretical truth.* The no miracle argument is a kind of meta-induction. (Psillos 1999, p. 79)[28]

The major (and well-known) difficulty with the type of argument proposed by Psillos is that of circularity. Indeed, the main premise asserts that it is reasonable to believe that particular theories are approximately true because they best explain the relevant phenomena. The no miracle argument would inherit its strength from the first-order level arguments, i.e. scientific abductions in favour of the reality of DNA, curved space-time, electrons etc. I do not take these abductive arguments to be convincing. But, even if, like Psillos, one thinks that they are, one immediately sees that this

[26] For a detailed critique of Putnam's argument, see Ghins (2002). A well-argued defence of the no miracle argument is developed by Mario Alai (2012).

[27] For Michael Devitt the no miracle argument is a bad abduction and proposes to stay at this first-order level. If a theory provides the best explanation of phenomena, then it is (partially) true. This is what he calls the "basic argument" for scientific realism (Devitt 1997, 2020). Devitt is certainly right in claiming that the basic argument is better than the no miracle argument since the former doesn't endow truth with an explanatory role. But this does not solve the problem of the reliability of the inference to the best explanation, since it is its very reliability that is at stake.

[28] Notice that Psillos' first-order level corresponds to what I called the second-order level. In my terminology, his version of the no miracle argument would be located at third-order level. For a discussion of various versions of the miracle argument see Ghins (2002).

premise presupposes the strength or reliability of the inference to the best explanation, whereas it is precisely this reliability that must be proved.[29] The no miracle argument can in no way be an induction based on the repetition of a supposed fact, namely that the application of inference to the best explanation, or of any other explanatory inference, succeeds in demonstrating that belief in theoretical truths is more rational. Indeed, Psillos and others must prove not only that it is rational to believe in some theoretical truths, but also that it is *more* rational to do so rather than to suspend our judgment. Such a proof cannot be delivered by appealing to inference to the best explanation.

Coming back to Putnam's formulation of the no miracle argument, it is worth pointing out that he is cautious enough to defend only the *partial*, not complete (in all details), truth of predictively correct theories. However, he doesn't propose any criterion to select the entities that are more reasonably believed to exist and to single out the claims regarding these entities that can be thought to be true, at least approximately. Besides, the no miracle argument is claimed to support belief in the existence of both detected entities (in the sense I understand it) and entities that are unobservable in principle. In order to make a sound case in favour of *selective* realism, most scientific realists have argued that the no miracle argument must be revamped to include criteria for identifying those parts of the theory that we can justifiably believe to be true.

In particular, many scientific realists assign great value to the capacity of a theory to predict novel unexpected observations. In such situation it is hard to object that the theory has been concocted in order to be able to account for known facts. Wouldn't be a miracle for a theory to successively predict bold observations that – given the background knowledge – are extremely improbable unless such a theory is at least partially true? It then remains to identify within the theory the parts which are indispensable to derive the novel unexpected facts and are thus very likely to be true. (Musgrave 1988; Psillos 1999; Leplin 1997; Alai 2014a, 2014b, 2017).[30]

Selective scientific realists argue for *local* or *topical* realism. In the first place, it is natural to ask that these parts play a decisive role in deriving correct predictions (Laudan and Leplin 1991, p. 462). With this first criterion we can discard belief in premises which are otiose, that is, not needed to yield empirical predictions. But this is far from being sufficient because the same predictions can be deduced from

[29] Psillos attempts to resolve the circularity problem by ingeniously resorting to the distinction between *premise circularity* and *rule-circularity* introduced by Braithwaite (1953). The no miracle argument would not be viciously circular since its conclusion is not contained in its premises. But it would be innocuously rule-circular, since it employs a rule of inference which is vindicated in its conclusion. (Psillos 1999, p. 82) But, even if the reliability of the inference to the best explanation is not explicitly stated in the *premises* of the meta-induction, it is certainly present in an implicit way. IBE is used, and then supposed to be reliable, in order to establish the reliability of the IBE. Vicious circularity is not dodged by Psillos' move. See also Douven (2021).

[30] I will criticise such recourse to novel facts to support scientific realism in the next chapter. Let me just remark here that far from supporting scientific realism, novel facts can be used by antirealists to show that they falsify well established theories pertaining to the background knowledge that were held to be true in virtue of their explanatory power...

distinct, mutually incompatible premises. This is a logical fact. Here we are confronted to the problem of the underdetermination of theories by observations, which will be discussed at length in the next chapter. To circumvent this difficulty, realists appeal to additional selection criteria. For example, they most often demand the presence of a causal connection between observations and the existence of particular entities (Chakravartty 2007, p. 60). It would however be ineffective to argue in favour of causal connections by appealing to their explanatory role, as I showed. On the other hand, if the presence of a causal link between direct observations and some entities that are not immediately observed could be empirically ascertained, we could trace back causes from the presence of their effects. This opens the way to a promising defence of the moderate form of epistemological scientific realism I favour, a defence which does not resort to the no miracle argument.

2.6 An Alternative to Inference to the Best Explanation[31]

In order to defend epistemological scientific realism, I will now propose an inductive strategy as an alternative to explanationist argumentations. As it turns out, when it comes to inferring the existence of entities that are detected, convincing abductive arguments draw their force from hidden *deductions which include causal propositions that are empirically and inductively justified.* Let me show this by looking at a simple example of inference to the best explanation, discussed by van Fraassen[32]:

1. Fact F: grey hair lies on the floor, cheese disappears, specific little noises are heard.
2. The presence of a mouse in the house (hypothesis H) *best explains* the facts F.

So, H is (likely) true: there is a mouse in the house.

According to this abductive argument, we have good reason to believe that a mouse lives in the house. However, the strength of this argument does not lie, despite appearances, in its abductive character. If it seems convincing, it is because it is supported by premises – established by *induction* – that are not explicitly stated, as I will now try to show. To prevent circularity, we must first define what a mouse is. A mouse is an animal identified by a set of specific observable properties. By definition, a mouse is an animal with four legs, a long tail, small ears, a pink snout. We have previously directly observed that animals which satisfy this definition,

[31] I am greatly indebted to an anonymous referee for detailed comments on this section.

[32] According to van Fraassen this argument leads to the conclusion that the hypothesis of the presence of a mouse is empirically adequate (1980, pp. 19–20). Thus, since a mouse is a directly observable entity, it is true that there is a mouse in the house. van Fraassen does not question the validity of the inference to a better explanation in the observable domain. I disagree with him (and many others) on this. I submit that inference to the best explanation is not truth-conducive, irrespective of the kinds of entities (directly observable, detectable, unobservable) it is applied to.

i.e. mice, also shed grey hair, eat cheese, make some typical noises etc. Such constant *association* between observed properties has been empirically ascertained by *induction*.

In addition – and this is the crucial point – constant *causal* correlations between some events have been observed. In general, causal connections can be ascertained by the empirical methods codified by John Stuart Mill (agreement, difference, concomitant variation) provided they are sufficiently refined.[33] The causal relation is asymmetrical since causes cause their effects and not the other way around. But when a causal relation is empirically ascertained, we have an if and only if logical relation between causes and effects: if the cause, then the effect, and conversely. Previously ascertained causal relations allow tracing back the existence of a cause from an instance of its effects.[34]

In the present example, the occurrence of grey hair on the ground and the disappearance of cheese regularly occur after the loss of grey hair and the ingestion of cheese by an animal. Repeated previous observations taught us that mice (defined as above), and *only* mice, have properties (as grey hair loss etc.) that are causally correlated, or connected, with the instantiation of directly observed properties (as the presence of grey hair on the floor, disappearance of cheese etc.) By *induction*, we conclude that in general the occurrence of the latter properties – the effects – imply the presence of a mouse, which is the cause. In this particular instance, from the observed presence of grey hair and disappearance of cheese we are then entitled to conclude – perhaps with dismay – that there is at least one mouse (the cause) in the house. We have then indirectly (since we made an inference) observed – and thus *detected* – the presence of something that has certain properties (has grey hair etc.). This is a detection, since we have not observed the mouse directly.[35] Later on, after a great deal of patience (or luck), we will (perhaps) directly observe a mouse, that is, a larger set of instantiated properties. This will obviously strengthen our belief in the presence of the mouse.

The abductive argument above is thus buttressed by the following deductive argument:

1. F: Facts

   Grey hair on the floor, disappearance of cheese, little specific noises.
2. C: Inductively confirmed causal relations

   Facts such as F imply the presence of animals that have grey hair, eat cheese etc. which cause these facts.
3. A: Inductively confirmed association

[33] I will not argue for this claim here. But notice that it is in line with my definition of a causal law (Chap. 1, § 7). Those laws must be inductively confirmed by observations and can be used to ascertain the presence of a causal link.

[34] See for example Psillos 2002, Chap. 1.

[35] Indirect observations by means of instruments such as telescopes or microscopes also rely on inferences since they presuppose the reliability of these instruments. They are thus detections, as we will see in the next chapter.

Mice, defined as four-legged animals etc. also have the associated properties of shedding grey hair etc.

Conclusion: a mouse is present.

This argument is deductively valid. But it is based on premises whose truth is established by *induction*, namely premises 2. and 3. Why do we *also* have – as a kind of *gift* – a lovely explanation of the observed facts? Because premise C describes *causal processes*! Losing hair, making noise and eating cheese are empirically attested causal processes, that is, sequences of properties that change in time. Even if we grant that the presence of a mouse best explains the clues, this does not imply that it is more rational to believe in its presence rather than to suspend our judgement. If we are entitled to conclude that a mouse is present, it is on the basis of previously attested causal processes which allow tracing back the existence of the cause from the presence of its effects. In turn, these causal processes ground the explanation of the empirical evidence or phenomena. The above argument in favour of the presence of a mouse is a logically valid deductive argument. If the premises F, C and A are true, this argument is also sound. Then, by making the conclusion a premise and taking F as a conclusion, we get a correct explanation of the factual evidence.

One could object that the above argument although valid is not sound because we must establish, in this particular instance, that the alleged causal connection obtains, i.e. that premise C is true. Some other cause could produce the observed facts. For example, it is possible that someone is playing tricks with me by putting grey hair on the floor etc. Such alternative causes are usually hypothesised by abductive reasoning. How can we eliminate possible competing causal explanations? I submit that we should rely on empirical evidence only and not on the degree of loveliness of explanations of such evidence. If there is no evidence that someone is playing tricks with me, then we have no reason to believe in this hypothesis. External empirical evidence is more reliable than the alleged internal virtues of competing explanations.

But what about *unconceived* possible alternative causes? (Stanford 2006) This is not a problem for inductivists because if these alternatives are unknown, *a fortiori* they are not empirically and inductively confirmed. The mere possibility of alternative causes that have not been envisaged does not prevent us from having *more* reasons, based on the available evidence, to believe in the presence of a mouse. Nevertheless, premise C is not established with certainty but only has a higher probability of being true.

Belief in the instantiation of detected properties is justified to the extent that we rely on deductive arguments whose premises describe causal connections, and which have been inductively confirmed. Notice that such confirmation can take place only if we are dealing with observable properties all along. In this case, the strength of arguments which have the logical form of abductions flows from the underlying presence of such causal connections, which allow tracing back causes from their effects. So, in fact, this kind of abductive arguments do not draw their strength from their explanatory power. They only *appear* to do so.

If one believes that explanatory power is indicative of truth, it is very tempting to enlarge the assumed reliability of inference to the best explanation to the realm of entities that instantiate purely theoretical – unobservable in principle – properties. Most scientific realists have succumbed to that temptation. Since such abductive arguments allegedly are by and large successful, why aren't we allowed to extend their reliability beyond the domain of possible observations by performing a new induction – a meta-induction, this time – that takes as its basis an ensemble of successful abductions? Insofar as we trust induction, isn't such an extension legitimate?

The answer is no. As I insisted, our task is philosophical. We are not dealing here with factual matters. We are not concerned with whether mice, electrons etc. exist, but with *assessing the strength of kinds of arguments* adduced for believing in their existence. And we are now evaluating the merits of a particular form of inference, namely inference to the best explanation or abduction. As shown above, the apparent strength of an inference to the best explanation rests on premises – confirmed by induction – that describe causal connections. On the basis of these causal connections, whose existence is empirically attested in the observable domain, it is more rational to believe that certain unobserved but observable causes exist. If we resort to descriptions of causal processes that have not been observed, we may well be able to construct a *lovely* explanation, but we have no good reason to believe that this explanation is correct. Why is that? Because when we postulate that purely theoretical unobservable entities are the causes of observed effects, it is impossible to ascertain, on the basis of observations, that there is a connection between causes and effects. In order to confirm that such a connection obtains, the causes, as well as their effects, must be observable – and also observed (directly or indirectly). This is possible with respect to the properties of mice and also properties of entities such as planets, molecules and electrons, as I will attempt to show in the next chapter. Lacking empirical evidence in favour of causal connections, inductive inferences to the truth of universal propositions that describe those connections is illegitimate. Thus, we have no good reasons to believe that explanations that refer to purely theoretical unobservable causes are correct.

2.7 Conclusion

In this chapter, I first showed that the "loss of reality" objection can be countered by adopting a theoretical attitude, distinct from the pragmatist attitude recommended van Fraassen. I argued that a model or structure, although abstract, can be claimed to be adequate to a concrete entity, because its representational capacity and correctness are based on facts described by true propositions. Granted, when I assert that a model is adequate to a phenomenon, it is tantamount to saying that it is adequate to the phenomenon *as represented by me*, as van Fraassen emphasises. It would be pragmatically incoherent to claim that a model is *not* adequate to a phenomenon and at the same time contend that it is adequate to the phenomenon as represented by me. Representing is always performed in the first person. There is no godlike point of view.

However, realism about observed entities does not stem from a pragmatic impossibility. Ordinary realism involves a conception of truth as correspondence between a proposition and an external independent fact, which makes this proposition true. It is always open to me to compare what I observe with a proposition that I affirm to be true in order to confirm or refute my affirmation.

I then proceeded to define scientific realism as a resolutely philosophical, normative and anti-naturalist thesis. Scientific realism is an epistemological thesis which can be briefly formulated thus: there are strong arguments that give us more reason to believe in the existence of entities that are not directly observed, but detected, and in the truth of selected statements about these entities. The difficulty with which the realist is confronted is that of characterizing these arguments and the grounds of their strength. I have shown that the famous no miracle argument for scientific realism is undermined by insuperable flaws. More generally, I have argued that the explanationist strategies promoted by most realists are doomed to failure.

In contrast to explanationist strategies, the alternative I have briefly presented in this chapter relies on causal generalisations, confirmed by induction. I will further develop this inductivist strategy in the next chapter with examples drawn from science. Such strategy is based on an analogy between, on the one hand, the reasons we have for believing in the existence of ordinary things and, on the other hand, the reasons that we have to believe that detected entities do exist. This way of proceeding complies with empiricist requirements since it consists in justifying our beliefs by relying on observations. The challenge to be met is to show that it is possible, given the satisfaction of certain conditions, to trace phenomena back to entities that cause them and to defend that we have better reasons to believe in their existence.

Chapter 3
Scientific Realism: A Defence

Abstract In the present chapter, I propose a way of defending epistemological scientific realism that differs from the usual explanationist strategies. Instead of developing a top-down explanatory argumentation which consists in evaluating the merits of several explanations of phenomena and then claim that the loveliest explanation is likely to be true, I offer a bottom-up inductivist strategy which starts from observations to go back to the causes that produce them, which can then be claimed to have been detected. I contend that this bottom-up approach is not vulnerable to the objections raised against inference to the best explanation (IBE). To illustrate this point I argue that the celebrated arguments for the existence of Neptune by Le Verrier and the reality of atoms by Perrin are convincing precisely because they implement such a bottom-up strategy. In addition, I attempt to show that this defence of scientific realism makes it possible to respond to the classical underdetermination objection according to which theories are undetermined by the available data. I also criticize the famous antirealist argument developed by Larry Laudan and known as the historical pessimistic meta-induction. I also show that the stability of reference of terms which denote detected entities can be preserved in spite of important theory changes. To do this, I adopt a causal descriptivist conception of reference based the works of Saul Kripke, David Lewis and Robert Nola and I apply it to the case of the electron.

In this chapter, I will present a way of defending epistemological scientific realism that differs from the usual explanationist strategies. Instead of developing a *top-down* explanatory argumentation which consists in evaluating the merits of several explanations of phenomena and then decree that the loveliest explanation is likely to be true, I will offer a *bottom-up* argument which starts from observations to go back to the causes that produce them. Although these causes are not observed, we have good reason to believe in their existence if they are detected, that is, causally linked to direct observations. This bottom-up approach has the advantage of not being vulnerable to the objections raised against inference to the best explanation. In order to illustrate this point I will argue that the famous arguments for the existence of Neptune by Le Verrier and the reality of atoms by Perrin are convincing precisely because they implement such a bottom-up strategy.

M. Ghins, *Scientific Realism and Laws of Nature: A Metaphysics of Causal Powers*,
Synthese Library 483, https://doi.org/10.1007/978-3-031-54227-5_3

In addition, I will attempt to show that this defence of scientific realism makes it possible to respond to the classical underdetermination objection according to which the available empirical evidence can always be accommodated by several incompatible theories. According to this objection, there is no cogent argument in favour of the truth of any of them. I will also criticize the famous antirealist argument developed by Larry Laudan and known as the *pessimistic meta-induction*. For Laudan, history of science reveals that most theories previously held to be true are now considered false. What guarantee do we have that the theories we now believe to be true will not be falsified tomorrow?

I will also try to show that the stability of reference of terms which denote detected properties can be preserved in spite of important theory changes. To do this, I will adopt a *causal descriptivist conception of reference*, inspired by the work of David Lewis and Robert Nola. Although the *words* or *terms* used to refer to unobserved – but detected – entities are certainly susceptible to undergo significant variations throughout history, we nevertheless have every reason to think that some terms used today denote the same entities as those formerly referred to by different words. For example, we will see that expressions such as "cause of magnetic light" or "molecular torrent" have been used to refer to what we today call "electron flow".

3.1 The Bottom-Up Strategy for Epistemological Scientific Realism

3.1.1 Observation

Empiricists are perfectly right to insist on the role of sensory experience when it comes to justifying beliefs. By starting from the analysis of the reasons we have for believing in the existence of observed properties of ordinary things, we will be able to highlight the distinctive features of sound arguments in favour of the existence of detected properties and the truth of propositions regarding the entities which instantiate these properties. As I will show, the reasons for believing in the reality of detected entities are analogous to those that justify our beliefs in the existence of ordinary things.

Why do I have more reasons to believe that there is a glass on my desk than not? Surely, it would not be irrational to believe that there is no glass at all, or to refrain from judging whether there is a glass or not. After all, philosophers who champion radical scepticism about the existence of the "external world" were far from being irrational. The world could be just a dream, a possibility that was seriously considered by Descartes.

How do I check that there is actually a glass on my desk? A glass is an entity characterised by a set of observable properties whose constant association I have been able to observe in many circumstances. A glass is hard, hollow, transparent, its edge is a closed line etc. I can make sure that a glass is present by directly perceiving some of its properties. In the context of the scientific realism debate, we saw that both realists and antirealists do not question the existence of ordinary things and implicitly adopt a form of direct realism.

Although I am aware that this claim will be challenged, I contend that properties like mass, charge, temperature etc. are observable. Some philosophers will be quick to rightly object that the *terms* "mass", "charge", "temperature" etc. belong to a theory-laden language. What is more, the meanings of these terms have only been clarified and stabilised after a long and painstaking historical process. This is also true. However, once the meaning of the term "inertial mass", for example, has been clearly fixed, we can, under favourable circumstances, check by direct observation *that* some bodies oppose unequal resistances when we attempt to modify their state of motion or rest. It is easy to verify that two balls at rest, made of the same material but of significantly different sizes, do not offer the same resistance to identical thrusts. The larger ball is more difficult to set in motion than the other: it has thus a larger inertial mass. In an analogous way, provided the meanings of "positive charge" and "negative charge" have been specified, it is possible to empirically ascertain the presence of charges of opposite signs, for example by directly observing (by vision or touch) the mutual attraction of bodies.[1] (Charges of opposite signs may very well have been hypothesised by abductive reasoning, the heuristic role of which I do not dispute.)

Some philosophers will promptly point out that there is a difference between observing property P and observing *that* something has the property P. Observing the property of being hard is not the same as observing *that* a glass is hard. Actually, this distinction matters little for the issue of realism. The truth of propositions and the instantiation of properties go hand in hand. Is it true that a hard thing is present on my desk? The truth of this statement depends on a fact, i.e. the instantiation of the property of being hard, attested in direct perception. Is it true that the inertial mass of ball A is larger than the inertial mass of ball B? Again, such assertion can be verified or falsified by means of simple observations. If it is true, ball A instantiates a larger inertial mass than ball B.

No property is cognitively accessible unless it is observable by us, either directly or indirectly. On this I agree with empiricists. However, among the observable properties I have included properties that are not usually held to be so by empiricists, such as charge and mass. These properties, like many others in science, can take various values, continuous or discontinuous (discrete). They are called *determinable* properties because they can take on determinate values. Certainly, due to the limitations of our senses, we do not have access, in direct perception, to very large or very small values of mass, charge, volume, velocity etc. However, since some values of these properties are observed, I will consider very large or small values (of velocity, for example) to be observable properties, in a broad sense, because they are indeed similar to directly observable properties: a very large velocity is still a velocity. Enlarging observability in such a way doesn't seem appropriate because, strictly speaking, a very large velocity is not observable. But this seems justified

[1] For more details, see Ghins 2017.

because resemblance ensures that we can have cognitive access to similar properties by means of detections whose reliability is justified by empirical inductions, as I will show later.[2]

With this condition of observability, it might seem that we are back to empirical adequacy, as understood by van Fraassen. This is not the case. For van Fraassen, what we see in a microscope does not support the belief in the reality of cells or mitochondria. According to him, only propositions that describe "appearances" can be judged to be true. Appearances here are the phenomena (images) *created by* the microscope, such as colour patches, shapes etc. that are immediately seen in this instrument.[3] Unlike Neptune or dinosaurs, to which direct observational access is precluded given our spatiotemporal location, mitochondria or viruses are beyond the direct perceptual abilities of humans. They can nevertheless be detected with a microscope. In keeping with his strict empiricist stance, van Fraassen recommends suspending our judgement as to the reality of such entities. I submit instead that a realist position as regards the existence of mitochondria or viruses can be justified, precisely because they have observable properties (dimensions etc.), in a broad sense of observability.

Let us now come to *unobservable*, *purely theoretical* or *undetectable properties*. This class contains the properties which escape or transcend any possible observation, even indirect, not only *in fact* but also *in principle*. Such properties would be empirically accessible only to beings possessing sensory modalities distinct from our five senses. Typically, we are unable to *imagine* purely theoretical properties because they do not resemble any property accessible in perception. These undetectable properties are common in the field of elementary particle physics. Think of (internal) spin,[4] strangeness, charm etc. Unlike properties such as volume, mass etc., it is impossible to ascertain their presence in ordinary sensory experience. As a consequence, we have no reason whatsoever to believe in the existence of such purely theoretical properties.

The issue of scientific realism hinges on the existence of properties that cannot be observed directly, due to practical constraints or the limitations of our senses, but which are detectable. The first condition of our being entitled to believe in the reality of such properties is that they are not only observable but actually observed. We are thus led to formulate the following observation condition.

Observation condition (O):

> *To have good reasons to believe in the existence of a property, it is necessary that this property be observed either directly or indirectly, that is to say, detected.*

---

[2] Notice incidentally that scientists often speak that way.

[3] van Fraassen (2008, pp. 100 ff). For a critique van Fraassen's position, see Bueno (2011) and Psillos (2014a).

[4] Even though the behaviour of the electron is in some respects analogous to that of a rotating charged body that is moving in an electromagnetic field, it is a mistake to imagine the electron as a sort of spinning top. This is why we take spin to be an *internal* property of the electron. On the other hand, it is defensible that magnetic moment is an observable property.

This condition is necessary but not sufficient: additional conditions are required for having better reasons to believe in the instantiation of a property. In normal circumstances, immediate observation is sufficient to warrant belief the instantiation of common properties such as hardness, but often the repetition of observations is required. For properties that are not immediately observed, justified belief in their reality must be supported by previously confirmed causal connections, as I shall now explain.

3.1.2 Causal Connection

A property is reliably detected on the basis of previously empirically ascertained causal relations with directly observed properties. To clarify things, it is useful to distinguish two kinds of detection. The first resorts to instruments in order to overcome the limitations of our senses or increase our perceptual abilities. Classical examples of sight enhancing devices include glasses, telescopes, microscopes... In this kind of access, which I call "homogeneous", the properties directly observed belong to the same modality (sight for example) as the properties detected and may even be identical to them. In a second kind of access, called "heterogeneous", the directly observed properties and the detected properties belong to distinct modalities. For example, by observing the visual properties of rays in discharge tubes J. J. Thomson was able to detect and measure the ratio between the charge and mass of the electron.

Let us start with the *homogeneous* kind of empirical access. For the sense of sight, which enjoys a privileged status in science, a great variety of detection instruments are available. First, we have the ordinary glasses used by people whose sight is slightly deficient. Who would claim that when a farsighted person observes a distant object with appropriate glasses, her observations are dubious on the pretext that they are not made directly, with the naked eye? But let us now examine what happens in the use of more powerful optical devices. In a telescope we see things, usually called "images", that have specific properties, for example a geometric shape. It turns out that geometric shape is also a directly observable property of celestial things such as planets. If we were close enough to the planet, we would right away observe that it is (approximately) spherical. This is why it is commonly said that we observe a planet through a telescope. In fact, what we directly observe is the properties of an image inside the telescope. The shape of the image A is related to the shape of the planet B by an if and only if relation. If A, then B, and conversely. Moreover, B causes A. According to my terminology, the shape of planet is *detected*.

These remarks can be extended to the various kinds of microscopes and telescopes which make entities such as viruses and distant galaxies visible to us. When immediate observations and those achieved with the aid of a microscope agree with each other, the microscope can be considered reliable, at least in the overlapping domain of observations. Then, by *induction*, we enlarge the domain of reliability of the microscope to properties that are not immediately visible. In addition, knowledge of the laws of optics, verified by induction in the domain of directly observed properties, entitle us to trust the microscope when dealing with properties invisible to the naked eye. Step by step, and by induction, we can extend the domain of accessibility of observable properties to increasingly larger domains of detection. This is why, as I said above, it is legitimate to consider very large or very small values of these properties as being observable in a broad sense even if these values are only detectable.

Notice that it is not always necessary to know the causal laws or the underlying mechanisms that govern the functioning of instruments in order to trust them. The ancient Romans used polished glasses to correct sight even though the laws of refraction, and *a fortiori* those of electromagnetism, were unknown to them. The same was true of Galileo and his contemporaries, who knew very little about the functioning of the *canocchiale*. Nevertheless, since close observations of the same object, a ship for example, and those made at a distance with the *canocchiale* were concordant, they could empirically ascertain the presence of a causal link between the properties directly observed inside the telescope[5] and the – detected – properties of a distant object. Even the Aristotelians fairly easily admitted that Galileo's *cannocchiale* was reliable.[6] This inductive way of proceeding, which Kitcher calls the "Galilean strategy" (2001, pp. 173–174), can be applied to other instruments.[7]

Of course, the knowledge of empirically verified relevant laws makes possible the fabrication of elaborate detection instruments, such as wire or drift chambers. With the help of these experimental devices we can detect and also measure the trajectories of charged elementary particles, like protons. In this case, to be entitled to claim that the lines directly observed on photographs or screens correspond to the trajectories of protons, we must rely on inductively confirmed causal laws that describe the interactions between protons and molecules. Typically, protons remove electrons from the molecules of a gas such as argon; these electrons are instantaneously captured by the neighbouring wires. The light emitted reveals proton trajectories.

In the *heterogeneous* kind of cognitive access, the detected properties do not belong to the same perceptual modality as the directly observed properties. For example, the properties that play a causal role in the production of light rays observed in cathode ray tubes, namely the properties of mass and charge, are not visual properties, like colour, shape etc. even though their presence can be ascertained by vision. Besides, J.J. Thomson was able to show that the light rays are produced by massive and charged particles emitted by the cathode, although he had no knowledge of the underlying (quantum) mechanism of such production. We know today that the particles emitted by the cathode interact with the molecules of the rarefied gas in the tube in such a way that light is emitted. The detected properties – charge and mass – of these particles and the directly observed visual

[5]Wesley Salmon (1984, pp. 232–235) similarly argues that it is legitimate to inductively extend the domains of reliability of telescopes and microscopes.

[6]The Aristotelians believed that the celestial world (from the sphere of the Moon onwards) was made of an incorruptible matter, distinct from that of the terrestrial world. Therefore, while accepting observational data, they tried to save Aristotle's physics by *ad hoc* assumptions. For example, they assumed that the Moon was enclosed in an immutable crystalline sphere.

[7]For the application of the same strategy to the microscope, see Magnus (2003).

properties of light rays belong to very different categories of properties. Nevertheless, Thomson managed to determine the value of the ratio of the mass and the charge of these particles – namely, the electrons – while being completely unaware of the mechanism for the production of these rays. To do this, he relied on the previously well-tested laws of classical mechanics and electromagnetism, as we shall show in some detail below.

Whatever the type of indirect epistemic access to properties, whether homogeneous or heterogeneous, we can rely on the presence of causal relations between these detected properties and actually directly observed properties. Previously inductively confirmed causal generalisations are crucial to attest the presence of causal relations. Knowledge of these relations entitle us to trace back the causes, the detected properties, from the effects, the directly observed properties. For sure, in ordinary experience what authorises us to justifiably assert the existence of a glass, a rose etc. is their actual presence in perception, in good observation conditions. But this perception is underpinned by causal connections, the details of which we do not know, but which we have every reason to believe exist, if only because when we intervene on the perceived object in a certain way, this is followed by systematic variations in the perception of its properties.[8]

We can check the trustworthiness of a new instrument by comparing the measurements it delivers with those obtained in the same empirical domain by an instrument the accuracy of which has been previously tested. If the results obtained by both instruments in such overlapping domain are concordant, we inductively extend the trustworthiness of the new instrument to wider detection domains, by applying the Kitcher's "Galilean strategy" I mentioned above. Step by step, we justify by this way of proceeding the considerable expansion of the range of detected properties following the invention of new instruments and measuring devices.

Moreover, if we are in a position to capitalise on detailed knowledge of the causal laws that describe the mechanisms at work in measuring instruments or observation devices, we have good reasons to believe that the causes of observed effects do possess some specific properties. These causes are instantiated properties (such as mass, charge or velocity) that are not directly observed, but which can be judged to have been detected. We then state a second condition that must be satisfied to entertain justified belief in the existence of detected properties.

Causality condition (C):

> *To have good reasons to believe in the existence of a property that is not directly observed, this property must have been detected, i.e. empirically ascertained to be causally connected by means of reliable instruments to properties that are directly observed.*

This condition is necessary, not sufficient. To ascertain the existence of a property, it is not enough to have detected it only once. Several detections are required, preferably using different methods of measurement, and these must give concordant results, as we will presently see.

---

[8] A similar point was made by Ian Hacking and Nancy Cartwright (Hacking 1983 and Cartwright 1983). See also Ruyant 2021.

3.1.3 Invariance

To make sure that a thing on my desk is hard, I touch it. I can also hit it and hear a sound which is typical of hard things. In addition, the entity has visual properties which indicate hardness. Three distinct *perceptual modalities* come into play here – touch, hearing and sight – which are mostly independent of each other. Each modality gives empirical cognitive access to the property of hardness. Within each modality, I can *repeat* observations in multiple ways and check whether their results are in agreement. By different ways of touching, I confirm that the property of hardness is instantiated. Similarly, by varying the ways of hitting, I hear each time a different sound, but which always reveals hardness. I am also free to apply the same methods to confirm the instantiation of other properties that reveal the presence of a glass rather than that of a vase or any other type of object. These observations indicate that an additional condition – the *condition of invariance* – must be satisfied to justify belief in the instantiation of immediately observed properties[9]:

Invariance condition (Ia):

To have good reasons to believe in the existence of a directly observable property, it is necessary and sufficient that repeated observations of that property by means of distinct and independent perceptual modalities give identical results, at least approximately.

For a directly observed property, the invariance condition is necessary and sufficient for justified belief in its instantiation. No doubt this condition of invariance is grounded in the truth of generalisations that describe Millian causal connections between the perceptions of a property (in favourable circumstances) and the actual instantiation of that property. We know that perception is a very complex process that involves causal connections (which are not fully known yet) between externally instantiated properties on the one hand and properties of our perceptual organs, our nervous system and our brain, on the other hand. Various observations, which rely on different causal connections but yield invariant results do increase our confidence in the reality of a particular property. Why? Because we learned from previous experience that by proceeding in this way, we run fewer risks of making mistakes. Previous experience has taught us, in a large number of cases, that beliefs supported by this procedure are less likely to be falsified. Indeed, when we want to dispel doubts about the attribution of properties to a particular entity, we repeat and diversify our observations. Again, such behaviour is justified by induction.

Notice, and this must be stressed, that our belief in the instantiation of a property (or set of properties) is *not* justified by arguing that this or these properties *best explain* the concordance between different perceptions. We do *not* have here an *inference* from various concordant observations. In each perception, the property is perceived directly. Repetition provides new perceptions of the same property, the agreement of which confers stability to our beliefs. Such stability takes place because we have learned by induction that the beliefs reinforced by this procedure are more resistant to possible falsification.

[9] See also Salmon (1984, p. 237) and Ghins (1992).

What can we say now about indirectly observed, that is, *detected* properties? We have seen that we had to empirically verify that causal connections obtain between directly perceived properties (clues) and detected properties to be justified in believing in the reality of the latter. But, by analogy with the invariance condition for immediately observed properties, we are led to require that repeated detections resorting to different methods deliver invariant results.

Invariance condition (Ib):

To have good reasons to believe in the existence of a detected property, it is necessary that repeated detections of this property by means of distinct and independent empirical methods give identical results, at least approximately.[10]

For detectable properties, this condition is necessary but not sufficient since the distinct empirical methods must also be reliable. This reliability is based on the requirement of causality presented above. Some empirical methods of detection use instruments, as in astronomy. Some do not, as in the detection of a mouse as we saw in the first chapter.

3.1.4 Measurement

In many scientific contexts it is impossible to accurately determine the value of a detectable property without the aid of *measuring* instruments. This is why we need to add a fourth condition: the *measurement condition.*

Measurement condition (M):

In the quantitative sciences, to have good reason to believe in the existence of a detectable property of a specific value, it is necessary for this property to be quantitatively measured by means of instruments whose reliability has been empirically established previously and independently.

This last condition makes it possible to satisfy the invariance condition for quantitative properties when a more precise agreement between data is required. Even assuming that mass is an observable property, its precise value cannot be determined without appropriate instruments. However, to ensure that measurement results agree – within the limits of observational error – they must be appropriately accurate. This measurement requirement taken in isolation is necessary but not sufficient: the measurements must also comply with the invariance condition...

[10] Such independent methods rely on distinct causal processes which make possible to detect the invariant cause (that is, instantiated properties) of several effects. This looks similar to what Descartes asserted in his 13 July 1638 letter to Morin: "But although there certainly are several effects to which it is easy to adapt diverse causes, one [cause] to each [effect], yet it is not so easy to adapt one identical cause to several different effects, unless it is the true cause from which they originate." (Descartes 1982, p. 104, footnote 42) However, Descartes' reasoning was explanationist while the invariance condition, as formulated above, requires that the reliability of each independent method be empirically and inductively ascertained.

3.1.5 The Four Requirements

We thus have four requirements or conditions for justified belief: Observation, Causality, Invariance, Measurement. For brevity, I will refer to these four conditions as OCIM conditions. The satisfaction of OCIM requirements – *taken together* – is a *necessary and sufficient condition* for having good reasons to believe in the instantiation of a *detectable* property. These conditions, and crucially the causality condition, allow ascending from directly observed properties to the properties that cause them. It is a *bottom-up* approach. On the other hand, there are no good reasons to believe in the existence of entities that possess only purely theoretical properties. Why? Quite simply because such properties are beyond any empirical cognitive access, or rather, for empiricists, are beyond cognitive access *tout court.* They are epistemically transcendent. In this case, it is impossible to empirically confirm that purely theoretical properties are causally connected to observed properties.[11] With regard to these purely theoretical properties, I recommend adopting an agnostic attitude. Such theoretical properties might exist, but we will never have strong arguments to believe in their reality.

We can now summarise these four conditions in the following way.

Requirement R:

> *To have good reasons to believe in the existence of a* detectable *property, it is necessary and sufficient that it has been measured a sufficient number of times by various methods whose reliability is grounded on empirically inductively confirmed causal connections between this instantiated detectable property and directly observed properties, and furthermore that the observation results or measurements obtained are concordant.*

This requirement is *normative*, in the sense that the satisfaction (or not) of the four OCIM conditions makes it possible to generally assess the strength (or weakness) of an argument offered in favour of the reality of some entity. The scientific realism I propose is thus a resolutely philosophical position. Its plausibility does in no way depend on whether scientists in fact mobilize (or not) arguments that comply with this requirement when they argue in favour the existence of specific entities. On the contrary, the cogency of this realist position rests on the similarity between the arguments adduced for believing in the existence of detectable entities and those in favour of belief in the reality of ordinary, immediately observable, entities. This realism is consistent with the view that science is an extension of common sense, as philosophers such as W. V. O. Quine maintained (Quine 1969, p. 129). The scientific properties that can be rightly believed to be instantiated are identical or similar to the properties of ordinary objects to which we have access in direct perception: these properties are all observable in a broad sense. Moreover, the existence of properties of entities postulated by a theory can be established provided their detection is confirmed by implementing procedures similar to those used in everyday life, that is, by repeating and diversifying our observations.

---

[11] According to special relativity, properties of entities outside our light cone are not accessible to observation. Suspension of judgement is then advisable. But some of these properties may become so over time.

The implementation of such inductive empirical strategy for defending selective scientific realism provides a welcome alternative to the vindications of scientific realism relying on the no miracle argument and explanationist strategies. I do not claim of course that the prediction of novel and unexpected facts doesn't have any relevance for reasonable belief in parts of a theory. But if the observation of a novel fact provides reason to believe in selected parts of a theory it is not because such observation generates psychological feelings of surprise or awe, which are devoid of epistemological force, or because the theory at stake has the resources to explain the novel fact. The epistemological force of novel observations flows from having ascertained convincing empirical evidence of a causal link between selected parts of the theory and the novel observations. This historically happened with respect to the standard textbook examples of the observation of a bright spot in the middle of the circular shadow projected by a circular screen (which provided evidence in favour of light waves) and the deflection of starlight in the vicinity of the sun (which supported Einstein's construal of the gravitational field in his general theory of relativity).[12] The importance of evidence in favour of a causal connection between novel observations and theoretical claims will be illustrated by discussing the simpler example of the discovery of Neptune in the next section.

3.2 The Case for the Existence of Neptune

In line with my opposition to naturalism, I do not think it legitimate to appeal to historical facts or history of science to support a philosophical position, whatever it may be. What is more, it is well known that the choice of historical episodes and the interpretation of texts that have come down to us may support distinct, even antagonistic, philosophical views. The historical empirical basis consists in traces (artefacts such as texts, tombs etc.) which is different in kind from the empirical basis provided by observation or experimentation in scientific contexts. True, empirical data are also open to interpretation, but they are facts that can be repeated, at least in principle. On the contrary historical facts are accessible only by means of traces and cannot be reproduced at will. Despite the difference between observational or experimental facts and historical facts, it may be illuminating to look at some emblematic episodes of the history of science which can serve as an *illustration* of the position defended here and to check whether the four OCIM conditions listed above are fulfilled by famous scientific arguments in favour of the existence of some detected entities. I thus propose to illustrate the bottom-up approach by first looking at the well-known argument in favour of the existence of the planet Neptune, which is frequently considered a paradigmatic example of inference to the best explanation.

[12] For a discussion of these two famous historical examples from an explanationist perspective, see Leplin (1997).

At the beginning of the nineteenth century, Bouvard had detected – with a telescope – that the trajectory of Uranus did not conform to the predictions of classical mechanics. Such anomalous observations are empirical evidence for an unexpected novel fact within the background knowledge of Newtonian physics. Several hypotheses were proposed to explain such anomalies: the influence of known planets or some comet, the presence of magnetic forces or an unknown planet, an incorrectness in the mathematical formulation of the gravitational force etc. But calculations based on Newton's laws showed that the detected perturbations of Uranus' trajectory are – so the argument goes – *best explained* by the presence of a hitherto unobserved planet. Hence, Adams and Le Verrier concluded that a hitherto unknown planet was the cause of the perturbations in the trajectory of Uranus. In 1846, such new planet, named "Neptune", was in fact detected by Johann Galle, not far from its predicted position. (Given the anomalous precession of the perihelion of Mercury, the same reasoning was applied by Le Verrier to conjecture the existence of a new hypothetical planet "Vulcan" between Mercury and the sun. This planet was never detected (Baum and Sheehan 1997). Later on, in 1915, the precession of the perihelion of Mercury was explained by Einstein's theory of general relativity. Newton's theory of gravitation was then shown to be empirically unsatisfactory for strong gravitational fields.)

I will now try to show that the strength of this abductive reasoning rests on a bottom-up argument, regardless of whether Adams and Le Verrier actually framed their argument in this way, which is a historical issue.

Planets are observationally defined as bright spots that, seen from the earth, move periodically along the constellations of the zodiac. They are also, according to Newtonian mechanics and the classical formula of the gravitational force, entities which revolve around the sun and have properties of mass, velocity and acceleration. Relying on the inductively confirmed causal laws of classical mechanics, we can infer from the effect detected with the telescope – namely the anomalies in the trajectory of Uranus – the existence of its cause, that is, the presence of a new planet whose motion conforms to classical mechanics. This approach is *bottom-up*, although the argument for the existence of Neptune is most often presented as a typical example of a *top-down* inference to the best explanation (see Douven 2021). As an alternative to explanationist reasoning, a sound deductive argument based on observation results and inductively verified laws can be framed thus:

1. F: Facts
   Anomalies in the trajectory of Uranus
2. C: Inductively confirmed causal correlations
   According to Newtonian theory, such anomalies imply the presence of a body with specific mass orbiting the sun along a definite trajectory, which is the cause of these facts.
3. A: Inductively verified association
   Planets, defined as bright spots moving periodically along the zodiac etc. also have a mass and obey the Newtonian theory.

Conclusion: a new planet, named "Neptune" exists.

By reconstructing the argument for the existence of Neptune in this way, we realise that it has the same form as the mouse argument discussed in the previous chapter. The detected deviations of the trajectory of Uranus are evidence for the existence of Neptune, in the same way as the observations of grey hair etc. were clues of the presence of a mouse. The force of the argument for the existence of Neptune does *not* lie in the assumption that its existence provides the best explanation of the anomalies in Uranus' orbit. Granted, we must assure ourselves that premise C is true. By abduction we are led to envisage alternative explanations of the detected anomalies. However, the elimination of alternative explanations is not accomplished by judging that their explanatory force is weaker, but by checking whether they are inductively supported by sufficient empirical evidence. Available competing alternatives to the existence of Neptune are discarded not because they are less lovely, but because they lack inductive observational support. Yet the truth of the crucial premise C is only likely because we cannot conclusively exclude the possible presence of some other unknown cause, even though we cannot obviously have good reasons to believe that such an unknown cause exists.

Abductive reasoning does help in imagining new hypotheses, but has heuristic value only. Abduction is not indicative of truth, as I contended. In order to ponder the credibility of a competing explanatory hypothesis, we must investigate whether observations warrant belief in alternative causes, such as the presence of a comet, by relying on inductively confirmed laws. When proceeding in this way we do not at all perform an abductive reasoning. If alternative hypotheses lack sufficient empirical support, there are discarded. Thus, the premise C is (probably) true and the deductive argument above for the existence of Neptune is sound. If this is correct, there were excellent reasons to believe in the existence of Neptune *before* its shape, colour, brightness etc. were detected. Of course, Johann Galle's later telescopic observations rightly reinforced belief in the existence of Neptune.

Obviously, the examples of the mouse and the planet Neptune differ in important respects. First of all, belief in the anomalies in Uranus' trajectory depends on the images seen in the telescope. Such belief is justified since the reliability of telescopes has been inductively established. For mice, we started instead from immediately observed properties and not from images. Besides, we had several kinds of clues: grey hair on the ground, disappearance of cheese... For Neptune, on the other hand, only the detected anomalies in the trajectory of Uranus were initially available. This is why, in order to dispel doubts about the existence of Neptune, the observations made by Galle were particularly important. But what is decisive for the justification of our beliefs, both in the case of Neptune and that of the mouse, is the previous empirical confirmation of the relevant causal connections, which ground the soundness of bottom-up deductive arguments.

To conclude this section, let us quickly verify that the argumentation for the existence of Neptune complies with the OCIM conditions and the requirement R stated above. First, all the properties involved are observable in a broad sense. Second, repeated observations of the Uranus anomalies with a reliable telescope yielded concordant measurement results. Both the measurement and invariance conditions are satisfied. Finally, by relying on these observations and Newton's laws, we ascertain that there is a causal connection between novel facts, namely the anomalies (the clues) and the presence of a new planet, with specific mass and trajectory: Neptune.

In order to further illustrate the normative function of requirement R in assessing the strength of a realist argument, I will now examine another famous historical episode in which the existence of entities more remote from immediate experience than Neptune was in question, namely the argumentation in favour of the reality of atoms elaborated by Jean Perrin.

3.3 Perrin's Argument for the Existence of Atoms

By the end of the nineteenth century, James Clerk Maxwell and Ludwig Boltzmann had developed the kinetic theory according to which gases are composed of particles whose motion is described by the laws of classical mechanics. This statistical theory provided a causal explanation of many properties of gases, as well as the ideal gas law. Moreover, in 1905, Albert Einstein relied on the kinetic theory to develop a beautiful explanation for the random movements of pollen grains suspended in a liquid, a phenomenon that had been observed (with microscopes) by the botanist Robert Brown as early as 1827. To explain Brownian motion, Einstein had assumed that liquids were, like gases, composed of corpuscles (indivisible atoms) in constant motion. According to this explanatory hypothesis, the random collisions of atoms (actually molecules) with pollen grains are the cause of their erratic agitation observed in microscopes. Significantly, and despite the undisputable explanatory power of the kinetic theory, of which the atomic hypothesis is a central claim, many renowned scientists like Mach, Duhem, Arrhenius, Ostwald and Poincaré, among others, remained sceptical. Of course, the vast majority of scientists conceded that the atomic hypothesis was practically fruitful for building explanatory and empirically successful theories. However, this practical utility was not (rightly) considered sufficient evidence in favour of its truth. Several objections were raised, in particular the lack of empirical evidence for a causal link between the hypothetical agitation of the corpuscles of the liquid and Brownian motion.

In 1888, when Léon Gouÿ proposed that the random motion the granules was caused by the motions of the particles of the liquid, he was guided by explanatory considerations typical of *abductive* reasoning. Abductive inferences to possible causal explanations are quite appropriate, but their role is purely heuristic, as I have argued. They do not dispense from subsequently checking, by inductive empirical methods, whether the supposed causal connection obtains or not. Indeed, such confirmation is required.

Various possible causes of granular motion were proposed and empirically ruled out. Gouÿ conducted experiments which excluded *external* causes such as the trepidations due to urban traffic, since granular agitation is also observed in quiet environments. Gouÿ's observations also showed that *internal* convection currents in the liquid could not be the cause of Brownian motion. Similarly, all the other proposed explanations were eliminated on the basis of inductive Millian methods.

As a consequence, the atomic hypothesis offered by Boltzmann, Einstein and others remained the *only* plausible explanation. (Psillos 2011b, p. 353).

However, the elimination of all the conceived explanations competing with the atomic hypothesis does not prove that the latter is true! To show that it is, one has to gather empirical evidence for the reality of a causal link between the motion of the granules and the collisions with the molecules of the liquid. This is what Jean Perrin did. Actually, he succeeded in proving that Brownian motion is evidence for its cause, that is, molecular agitation. I will now propose a *reconstruction* of the argumentation in favour of the existence of atoms which is inspired by Perrin's paper *Mouvement brownien et réalité moléculaire* (1909b). To make things clear right away, I will not claim that Perrin in fact argued in accordance with this reconstruction. My main claim is not historical, but epistemological. I will attempt to show that Perrin's observations do justify belief in a causal link between Brownian motion and molecular agitation, not because the latter best explains the former, but because the former is conclusive evidence for the latter, in conformity with a bottom-up argumentation.

In his 1909 paper, Perrin (1909b) begins by recalling that previous observations resulted in the elimination of all available explanations of Brownian motion, except the atomic hypothesis (or more accurately, the conjectured properties of atoms that classical mechanics and the kinetic theory attribute to them; this is why the expression "atomic theory", sometimes used by Perrin, is more appropriate). However, Perrin appears to be wary of an explanationist strategy when he says:

"Instead of taking this hypothesis ready-made and seeing how it renders *account of*[13] the Brownian movement, it appears preferable to me to show that, possibly, it is *logically suggested* by this phenomenon alone, and this is what I propose to try." (Perrin 1910, p. 7).

After arguing in favour of the plausibility of the atomic hypothesis from the observed characteristics of Brownian motion,[14] Perrin reviews and spells out some consequences of the atomic hypothesis, such as Avogadro's hypothesis and the principle of equipartition of energy for gases and fluids. The Avogadro hypothesis, which is suggested by the rules of substitution of elements in chemical reactions, states that a gram-molecule of any substance contains the same number N of molecules. N is Avogadro's number. According to the principle of equipartition of energy for gases, the mean kinetic energy of molecules is identical for all gases at the same temperature. Van 't Hoff's law allows us to extend this principle to liquids.

> At the same temperature all the molecules of all fluids have the same mean kinetic energy, which is proportional to absolute temperature. (Perrin 1910, p. 19).

Perrin then conjectures that the principle of equipartition can be extended to Brownian particles: the granules suspended in a liquid have the same mean kinetic energy as the molecules of the intergranular fluid. This is Perrin's main assumption,

---

[13]The italics in this quote are mine.

[14]Especially, the occurrence of an equilibrium between the coordination and the de-coordination of the motions of the granules.

whose confirmation or falsification would support or doom the kinetic theory, which specifies what the properties of the conjectured atoms are and the nature of the hypothetical causal link between granular motion and the random collisions with the atoms (molecules) of the intergranular fluid.

> (…) if we find a means of calculating granular energy in terms of measurable magnitudes, we shall have at the same time a means of proving our theory. (Perrin 1910, p. 21).

According to the kinetic theory, the mean kinetic energy W (equal to $m <v>^2/2$, where $<v>$ is the mean velocity) of a molecule is given by Boltzmann's formula:

$$W = (2/3)R/NT (R \text{ is the ideal gas constant})$$

This formula is part and parcel of the kinetic theory. In virtue of the reasoning above, the granules of the emulsion satisfy this formula if and only if the atomic theory is true. The challenge now is to devise a way of determining the energy of the granules on the basis of observable quantities. Achieving this would be tantamount to confirming the existence of the mentioned causal link as well as the atomic hypothesis, provided the value found corresponds to previous rough estimates of molecular energy described at the beginning of the paper.[15] Such confirmation would be reinforced by the concordance with the values of the mean energy of molecules obtained by more independent methods.

Perrin was an exceptionally gifted experimenter. He managed to produce very small grains of mastic and gamboge with approximately the *same size*. Those grains are suspended in an emulsion which is in a state of statistical equilibrium, that is to say, the number of grains at a given height in the container remains on average constant.

Supposing that the granules behave like a gas, Perrin obtained a mathematical formula in which the granule distribution exponentially decreases with height, just as the concentration of air in the atmosphere decreases with altitude according to Laplace's exponential law. The concentration of granules is a function of their height h, average kinetic energy W^*, volume φ of a grain, grain density Δ, liquid density δ and the gravitational acceleration g. Concentration of the grains is n at height h and n_O at the bottom of the container. (Perrin 1910 § 14, Psillos 2011b).

$$\log(n_0/n) = (3/2)W^*(\Delta - \delta)\varphi gh$$

Observation confirms that the distribution of grains is described by an exponential function. We have thus an inductive confirmation of the assumption that the grains of the emulsion behave as a gas. The figures below represent what Perrin saw in a microscope. The figure on the left shows that the concentration of the granules decreases with height. The figure on the right pictures shows the distributions of

[15] Especially, van der Waals' estimated values of molecular energy and Avogadro's number ($N \approx 45 \times 10^{22}$). (Perrin 1910, p. 17)

granules at heights which correspond to the horizontal lines in the figure on the left. (Perrin 1909b).

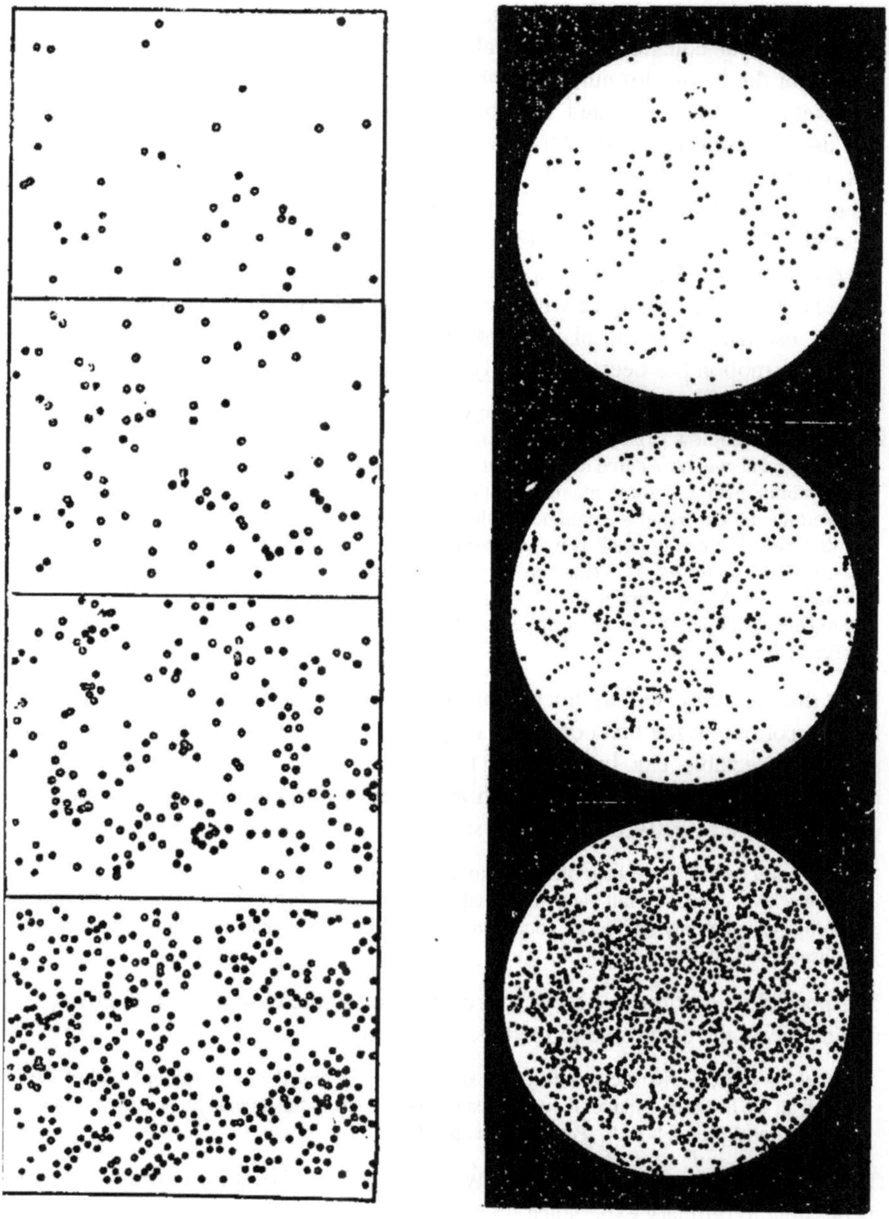

In Perrin's formula above, *all quantities are measurable except the mean kinetic energy W* of the granules*. The value *W** can then be calculated on the basis of properties measured by observations. Perrin concludes that "(...) if the value it indicates for W^* is the same as that which has been approximately assigned to the molecular energy (...) the origin of the Brownian motion will be established (...)" (1910, p. 24).

If we substitute W^* with Boltzmann's formula for the mean kinetic energy, we get a mathematical equation in which all quantities are measurable and detectable, except for the Avogadro number *N*, the value of which can now be determined by observations. Perrin repeated his observations with several kinds of grains and liquids. The finite values of *N* thus obtained were concordant. Perrin was then in a position to conclude – and this is the crucial point – that a supposed property of the alleged cause, namely that it is composed of countable elements, is empirically determined by observed properties of its effect. Just as the grains are discrete, the intergranular liquid is also composed of moving particles (atoms). Moreover the values obtained for W^* agree with previously estimated values, in particular van der Waals' estimate. The causal link between the agitation of liquid molecules and Brownian motion has been empirically ascertained.

> I do not think this agreement can leave any doubt as to the origin of the Brownian movement. To understand how striking this result is, it is necessary to reflect that, before this experiment, no one will have dared to assert that the fall of concentration would not be negligible in the minute height of some microns, or that, on the contrary, no one would have dared to assert that all the granules would not finally arrive at the immediate vicinity of the bottom of the vessel. The first hypothesis would lead to the value *zero* for *N*, while the second would lead to the value *infinity*. That, in the immense interval which *a priori* possible seems possible for *N*, the number found should fall precisely on a value so near to the value predicted [van der Waals' estimate] certainly cannot be considered as the result of chance. (Perrin 1910, p. 46).

Does this reconstructed argumentation fulfil the four OCIM conditions stated earlier? According to the first condition, only properties that are either directly observable or detectable can be believed to be real. We saw that the properties of countability and kinetic energy which are attributed to atoms, are detectable properties. As soon as in 1903 Perrin had said:

> It seems to me that we have the right to attribute more reality to molecules, atoms and corpuscles (...) I do not fall back into metaphysics. I do not forget that sensation is the only reality. It is the only reality, on condition that we add to the actual sensations all *possible* sensations.[16] (Perrin 1903, p. ix).

The possible sensations are those that correspond to the observable properties, and, among these, the detectable properties. As Perrin says:

> (...) these particles [mastic granules etc.] simply serve to reveal an internal agitation of the fluid, all the better the smaller they are, just as a cork follows the movements of the waves of the sea better than a boat.[17] (Perrin 1909a, p. 353)

This quotation expresses in a particularly vivid way Perrin's concern to remain as close as possible to ordinary experience. In his 1913 bestseller book *Les atomes* he states:

[16] My translation.

[17] My translation.

> *The objective reality of the molecules therefore becomes hard to deny*. At the same time, molecular movement has been made visible. The Brownian movement is a faithful reflection of it, or, better, it is a molecular movement in itself, in the same sense that the infra-red is still light. (Perrin 1916, p. 105).

Thus, Perrin has managed to show that molecular motion has been *detected* by means of the motion of the granules. According to the observability condition (Sect. 3.1.1), the entities that are conjectured to exist must possess observable properties, in the broad sense. In this case, we are entitled to apply previously inductively confirmed laws, such as the laws of classical mechanics, to these hypothetical entities. The properties of the atoms (molecules) such as number and energy are observable, since they are also properties observed for ordinary objects, although for molecules these properties are detected only. From an empiricist perspective, which is the one adopted by Perrin, our "ontological stock" or "furniture of the world", to use expressions which are common in the philosophical literature on scientific realism, can only contain entities that possess properties sufficiently close to ordinary sensible experience.

Perrin also insisted that his determination of the Avogadro number is more direct since it is deduced from observations. His argumentation is convincing because it starts from the effect (Brownian motion) to justify belief in the reality of its supposed cause (molecular agitation). Such argumentation is bottom-up rather than top-down. As Perrin says:

> (...) the microscopic observation of colloids and suspensions had familiarised me with Brownian motion, and soon gave me the possibility of a direct quantitative verification of the molecular theories (...)[18] (Perrin 1923, p. 29)

From the values of observed quantities, together with Boltzmann's formula, the value of Avogadro's number is directly deduced.

One could object that the whole argumentation is developed within the framework of the kinetic theory whose truth is at stake. This is correct. In order to inductively confer the presence of a causal link, we must be precise of what we mean by the conjectured cause in the first place. Since the motion of the molecules of the liquid was not visible, only their kinetic energy – which depends on the temperature and Avogadro's number – can be epistemically accessible. Then, the calculation of Avogadro's number by means of the kinetic theory (Boltzmann's formula) together with repeated observations give a value which corresponds to the value estimated by, among others, van der Waals. What we have here, is a typical inductive confirmation of a theory and causal link which had been initially conjectured by abduction.

As for the invariance condition, it is first of all satisfied by the concordance of the values of N obtained from Perrin's observations of the distributions of the granules in different types of emulsion. What is more, from the observation of Brownian motion, Perrin obtained four sets of values for N by applying independent methods (grain height distribution, displacements, rotations, diffusion) (Perrin 1916, p. 206). The

---

[18] My translation.

agreement of these values of N already provided a good confirmation of the atomic hypothesis.

In addition, the value of N deduced from Perrin's experiments was close to the values measured by *other independent methods*, such as those based on the viscosity of gases, radioactivity and black body radiation (Nye 1972, pp. 131–136). In this regard, Perrin speaks of the "miracle of concordances" (Perrin 1916, p. 206) between the values of Avogadro's number obtained by 13 independent methods. The value obtained by his method of "distribution of grains" in Brownian motion, that is, 68.3 $\times 10^{22}$, is in very good agreement with the other results. Avogadro's number is thus an *invariant*, a number that does not vary significantly according to the different measurement processes, in conformity with the third condition. In view of this agreement, the assumption that the grains behave like large molecules, which is equivalent to the assumption that W^* is equal to W, is very unlikely to be false.

The use by Perrin of the word "miracle" should not mislead us to believe that we have here an instance of abduction, or even the no miracle argument! The concordance of the values obtained by different methods simply enlarges the inductive basis in favour of belief in the atomic hypothesis, and indeed the correctness of the hypothesised (by abduction) explanation of Brownian motion. Remember that when I discussed the invariance condition, I pointed out that we have inductively confirmed that beliefs justified by the concordance of a large variety of observations were less likely to be revised.

Such inductive confirmation of the atomic hypothesis as construed by the kinetic theory does not stop here. With Avogadro's number we can calculate the masses and volumes of molecules. Again, these values are consistent with results obtained using several independent measurement methods. We could also take these coincidences to be new empirical successes or novel facts which provide additional inductive confirmation of the atomic theory. This translates into an even wider enlargement of the inductive support of the kinetic theory and, as a consequence, the causal link between the motions of liquid molecules and granular movements.

It then seems difficult to defend that Perrin's argument is explanationist and top-down rather than inductively based and bottom-up. True, one cannot deny that the atomic hypothesis does explain Brownian motion and the concordance between the measured values of Avogadro's number as well as between the measurements of certain properties of molecules, such as their dimensions and masses. However, as I said above, once we have managed to empirically verify the presence of the cause, we can implement a top-down argumentation and explain the available observations, and even predict new ones. As Perrin contends, "to explain the complications of the visible in terms of invisible simplicity" (1916, p. VII) is certainly one of the objectives pursued by scientific activity.

Yet the alleged simplicity of an explanation cannot justify belief in its truth. Such explanatory hypotheses can play a heuristic role by suggesting plausible causes whose reality must then be empirically confirmed, in compliance with the bottom-up approach. Now, in the case of Perrin's argument, this is indeed what we have. From the observation of Brownian motion, Perrin proposed a possible cause: the molecular agitation (that is, the existence of atoms) as described by the kinetic theory. It must also be assumed that the hypothetical cause possesses certain internal detectable – mechanical – properties in order to be able to verify that the supposed cause in fact produces the movement of the grains. No circularity creeps in. Perrin showed that the reality of the proposed cause is inductively confirmed by his repeated observations of Brownian motion and other kinds of observations.

Finally, I have no difficulty in conceding that Perrin's texts also contain explanatory reasonings. I do not claim that Perrin exclusively implemented a bottom-up argumentative strategy. Nevertheless, I argue that the strength of his argument – its backbone – lies in the verification of the reality of causal connections between properties detected with the microscope and the properties of the entities that cause them.[19] This is why his argument is forceful and supports our belief in the discontinuity of matter. My point is normative. Studying the reasons why scientists converted *de facto to* the atomic hypothesis on a massive scale at the beginning of the twentieth century is a task for historians and sociologists of science, but not for philosophers.

3.4 Two Objections

Before turning to the antirealist objections arising from the underdetermination of theories by data and the pessimistic meta-induction, I would like to address two difficulties with the realist position defended above. The first concerns the invariance condition. Doesn't this condition boil down to a disguised appeal to inference to the best explanation? Wouldn't it be a miracle or a very unlikely coincidence if the concordance of a large quantity and variety of observations or measurement results were not caused by an entity with appropriate properties? First, I do not advocate an epistemology based on mental, subjective entities such as sensations or *sense data* (Ghins 2000). If this were the case, one could legitimately ask whether the belief in an entity is be justified given that it provides the best explanation of the concordance between sense data. As said before, I embrace a form of direct realism, according to which we immediately perceive real properties. Second, if we observe a property directly, we do not infer, even implicitly, the existence of that property from our observations. If there is any inference, it is inductive. In fact, we have frequently observed that the repetition of consistent observations protects us better against the risk of error, that is, the risk that our beliefs be falsified. The invariance condition is supported by a meta-induction.

---

[19] I thus disagree with Achinstein, who proposes a reconstruction of Perrin's argumentation as based on an "explanatory connection" between the concordance of *N* measurements and the atomic hypothesis. (2001, pp. 259–260) For a probabilistic reconstruction of Perrin's argument, which starts from the observation of the concordance of the value of N to infer the probability of the truth of the atomic hypothesis, see Psillos (2011b). Against it, Bas van Fraassen (2009) claims that Perrin's approach does not establish the truth of the atomic hypothesis, but only makes possible to empirically ground the values of theoretical quantities, such as Avogadro's number.

Also, we saw that in the case of detected properties – like those of a planet, a molecule etc. – the available observations allow us to deduce their existence from previously verified causal connections. No doubt, in this kind of situation, these properties causally explain the observed evidence. However, we do not infer the existence of properties because they deliver the explanation that seems the best among a set of possible explanations we have managed to concoct. Within the framework of our causal approach, we are not looking for possible hypotheses from which we can derive, in a top-down approach, an explanation of the appearances. Instead, we try to climb the ladder of causal connections by taking the observed effects as a starting point. It is the proven existence of a causal connection that is an indicator of truth, not the supposed beauty of an explanation. When we have succeeded in empirically confirming the reality of a causal connection, we are immediately rewarded, as a kind of *bonus,* with a correct causal explanation of the phenomena. Moreover, if the mechanism of the causal connection is known, we get in addition a mechanistic explanation, in the sense specified in Chap. 1.

This brings us to the second difficulty, which has to do with the causality condition. For the enthusiasts of inference to the best explanation, belief in the existence of Neptune, atoms and electrons (see below) are considered classic examples of the truth-conduciveness of abductive reasoning (Douven 2021). Can we not imagine a possible causal connection between the deviation of Uranus' orbit and the presence of a comet? Couldn't we hypothesise an influence of the magnetic field of Jupiter on that of Uranus, as indeed some people did at that time? We would then have several possible causal explanations, among which the best (by some standards) would probably be the correct one. As Peter Lipton points out:

> Inference to the Best Explanation does not require that we infer only one explanation from the data, but that we infer only one of *competing* explanations. (Lipton 2004, p. 62)

I do not deny, of course, that various plausible causal connections can – and even should – be considered. But I claim that some possible causal connections should be discarded not because they are less beautiful, but rather because they do not conform to the observations.

Finally, remember that the no miracle argument, which is a special case of abduction, does not involve any causal considerations. Truth is not a physical cause. Yet according to the bottom-up approach, observational assessment of the occurrence of causal relations plays a crucial role in the defence of the selective realism I propose. Now, even if we limit ourselves to causal explanations, how is it possible for us to evaluate their respective merits? What *a priori* criteria should be used? To rescue the truth-conduciveness of abductive reasoning, it would be moot to decree that the best explanation is the explanation which rests on a empirically verified causal relationship. In that case, we would no longer have abduction, but inductive reasoning. Of course, I grant that the explanation based on empirically attested causal relations is the best explanation. However, the reason that this explanation is the best is that it is grounded on observations and laws inductively verified *a posteriori*.

Even if a consensus were in place regarding the features which would allow us to rank explanations according to their loveliness, the challenge facing the realist who adopts an explanationist strategy is to "show how judgements of likeness [of truth] are determined, at least in part, by explanatory considerations" (Lipton 2004, p. 61). On this, I fully agree with van Fraassen (2017) that we have no reason to believe that *a priori* considerations on the merits of an explanation plead in favour of its soundness, that is, the truth of its premises.

3.5 The Antirealist Argument from the Underdetermination of Theories by Observations: A Critique

The underdetermination argument is arguably the most powerful argument against scientific realism. Given an observationally correct theory, it is in principle always possible to construct another incompatible theory that equally saves the phenomena. In other words, it is always possible to construct a plurality of theories, incompatible with each other, all of which are consistent with the same observational or measurement results. Observations are then unable to determine which of these empirically equivalent theories can be justifiably believed to be true (we are talking here about an equivalence with respect to the observations actually available; this *weak* equivalence is distinct from *strong* empirical equivalence which concerns all possible observations, including future ones). In this sense, theories are underdetermined by phenomena. Observations do not provide better reasons to believe in the truth of a theory rather than another among a set of (weakly) empirically equivalent but incompatible theories. Future observations may well discard some theories in competition, but the realist is interested in the criteria which would legitimate belief in a theory at a particular moment of time, that is, now![20]

The underdetermination thesis goes back to Plato's *Timaeus* (29d). Even if they save the phenomena, our theories are no more than "plausible myths". These could be true, of course, but we are not in a position to justifiably believe that they are. The underdetermination thesis was taken over in 1644 by Descartes in his *Principles of Philosophy* (Descartes 1982, p. 106), and much discussed in the context of the debate between scientific realists and antirealists until today.

In a certain sense, this thesis is banal and therefore difficult to contest. Indeed, no one can *a priori* exclude the possibility that in the future scientists will produce a new (interesting and not artificial) theory that is observationally equivalent to a theory accepted today, but which partially contradicts it. The weakness of the underdetermination thesis lies precisely in its banality. In practice, in order to challenge a realist position with respect to a particular theory, the antirealist sceptic

[20] In the rather rare cases of strong empirical equivalence which are limited to physics and remain controversial (see Ghins 2005), suspension of judgment is in order.

cannot rest content with waving the red flag of the mere *possibility* of empirically equivalent alternative theories. It is up to the sceptic to *demonstrate* that it is *always* possible to construct theories which are empirically equivalent and incompatible to a theory accepted at some time.

It must be granted to the antirealist that the mere possibility of observationally equivalent theories is easy to demonstrate, whether one adopts the syntactic or the semantic approach. To start with the syntactic approach, for which a theory is primarily a set of propositions, we can readily point out that the same observational proposition (or set of propositions) can be deduced, by means of valid reasoning, from various incompatible sets of premises. This is a logical fact. On the other hand, an antirealist who adopts the semantic point of view, will immediately recall that distinct and irreconcilable theoretical superstructures can contain the same empirical substructures. This is a direct consequence, in the semantic approach, of the requirement of set inclusion, a requirement which is little restrictive and allows much leeway. Realists are aware of the threat of underdetermination. They concede that it is illusory to defend that the mere consideration of the empirical consequences of a theory is sufficient to pronounce that the theory is true. Empirical correctness *alone* does not warrant justified belief in the truth – even partial – of a theory. Additional conditions must be satisfied to have good reasons to believe that a predictively correct theory is partially and approximately true.

For the partisans of explanationist strategies, the explanatory power of a theory counts as a reliable indicator of its (at least partial) truth. A believable theory doesn't just save the phenomena; it must also provide a satisfactory explanation of them. It then becomes difficult to demonstrate in general that it is always possible in principle to construct alternative theories whose explanations would be as lovely (or even more so) as the explanations provided of an available theory. Taking explanatory power into consideration would then deliver, the explanationist argues, a selection criterion which would provide a bulwark against the underdetermination objection. But, as I argued in Chap. 2, there is no *a priori* guarantee that reality is obliging enough to comply with our explanationist demands. Realists who invoke the greater explanatory power of a theory to claim that such theory is more likely to be at least partially and approximately true are unable to overcome Voltaire's objection.

I acknowledge that an empiricist, even moderate, doesn't have a conclusive rejoinder to the underdetermination objection. However, I submit that the implementation of a bottom-up strategy enables us to undermine its potential devastating consequences for the realist position. Even if we have managed to confirm – on the basis of observations – the presence of a causal link between a conjectured cause and its observational effects, we are not in a position to exclude that some alternative cause is in fact responsible for the observations, especially when the procedure is intricate, as we saw in the example of Perrin's argumentation in favour of the existence of atoms. Nevertheless, attesting the presence of a causal link relies on empirically confirmed causal generalisations or laws that we have good reasons to believe to be true. Therefore, justified belief in the reality of detected causes is grounded on a selection of relevant observations, precisely those which have a bearing on the sought-after inductive confirmation. The case in favour of the reality of entities posited by a theory is then grounded on qualified observational material, and not on the global consideration of all the empirical consequences of the theory. If my argumentation is correct, such a way of confirming a causal connection between selected observations and some theoretical entities, dodges underdetermination by giving more reasons to believe in the partial truth of a theory. Of course, all this does not preclude that novel observations may lead to the falsification of thus justified theoretical claims in the future.

3.6 Larry Laudan's Pessimistic Meta-Induction: A Critique

No version of epistemological scientific realism could avoid responding to the notorious antirealist argument proposed by Larry Laudan in 1981 and which has not ceased to foster interest and controversy since then. According to Laudan, reviewing the history of science compels us to recognise that central theoretical claims of once empirically correct theories held to be true by scientists by the vast majority scientists at some time in history have now been acknowledged to be false. As a result of new discoveries, scientists have been led to recognise that the *central explanatory terms* of theories previously believed to be true did not refer to any existing entity or, in other words, were devoid of any referent. The famous "Laudan list" (1981, p. 32) includes, among others:

> The crystalline spheres of ancient and medieval astronomy
> the humoral theory of medicine
> 'catastrophist' geology, with its commitment to a universal (Noachian) deluge
> the phlogiston theory of chemistry
> the caloric theory of heat
> the vital force theories of physiology
> the electromagnetic ether
> the optical ether
> theories of spontaneous generation

Laudan concludes that our present beliefs in the existence of unobservable entities postulated by our best theories will be challenged sooner or later and eventually abandoned, as has befallen to past theories about crystalline spheres, ether, phlogiston etc. In the future, it is highly likely that the scientific community will no longer endorse the existence of viruses, electrons etc. Prudence advises us to embrace an agnostic attitude towards the existence of viruses and electrons as described by current theories.

Laudan deploys an inductive argument based on sociological facts of acceptance and subsequent rejection of scientific theories throughout history. More precisely, his argument is a *meta-induction*, often referred to as a "pessimistic meta-induction", since its inductive basis is a set of scientific theories that have been considered to be true for a certain period of time only to be later recognised as false. This inductive basis certainly is empirical, but it is historical and sociological in kind. It is thus situated on a different level than that of the observational predictions of a specific theory. No wonder then that Laudan gives very few details about the contents and the empirical correctness of the theories he mentions. He merely takes as his inductive basis a set of theories believed to be true for some time by the scientific community, without addressing the main philosophical issue, namely the normative reasons that were supposed to justify epistemic commitment to these theories. In my view, the main question an antirealist philosopher should be concerned with is the following: how cogent were the reasons for believing in the reality of the entities listed by Laudan? In particular, were these entities – with the properties attributed to them by these theories – connected to the observations by empirically confirmed causal links? It seems that they were not.

If we take a closer look at the theories mentioned in Laudan's list, we realise that there is no sufficient evidence of a causal connection between a mentioned entity and observations. Take the famous crystalline spheres example. These transparent and hard spheres were postulated by Ptolemy in the second century AD with the purpose of *explaining* observed facts, namely the stability of the cosmos and the regularity of the planetary trajectories over time. These spheres are observable in principle because their property of hardness (among others) is. True, the theory could predict the motions and positions of the planets with acceptable accuracy, but no causal connection between the crystalline spheres and the trajectories of planets had been empirically attested. There was no way to detect the supposed hardness of these spheres with the help of available trustworthy instruments or empirically established laws. In other words, those spheres had not been detected. Moreover, in Ptolemaic cosmology crystalline spheres were supposed to interpenetrate. This shows that for Ptolemy the notion of hardness did not fully correspond to the usual meaning.

Let me emphasise that the targets of Laudan's critique are the denotations of terms that play a central *explanatory* role, that is, terms that are involved in most explanations of the phenomena belonging to the empirical domain of a given theory. In doing so, Laudan wages a frontal attack against the top-down strategy mobilised by most realists. And he fully succeeds. Any attempt to rescue realism while remaining faithful to explanationist strategies is futile. Such strategies are not indicative of truth as we saw in Chap. 2. Indeed, the strategy generally marshalled by realists to counter Laudan amounts to stating more restrictive criteria to which an explanation should comply in order to be satisfactory and thus to have more chances of being correct. These criteria aim to constrain the admissibility of an explanation by requiring, for example, that the central terms play an essential role in the explanation of empirical predictions. (Psillos 1999, pp. 104–105).[21]

In fact, as noticed by Timothy Lyons (2002), Laudan's argument can be framed as a *deductive* argument in which the crucial premise is established by his meta-induction. Indeed, all the theories believed to be true in the past were observationally

[21] Such strategy is not exempt of risk, however. Demanding ever more stringent criteria for belief in unobservable entities can lead to a minimal form of realism that becomes too meagre to deserve to be still called "scientific realism" (Cordero 2017). For a critique of Psillos' *divide et impera* strategy, see Vickers (2013).

correct (and thus successful in Putnam's sense). Moreover, some of them even managed to predict novel facts. Yet most of them were later falsified. We thus have a deductive argument with a *modus tollens* form, which goes like this:

1. If scientific realism is correct (A), then the central explanatory terms of successful theories have a referent (B).
2. Historical evidence shows that the central explanatory terms of once successful theories have no referent (not B).

   Conclusion: scientific realism is not correct (not -A).

For Laudan, premise 2 is established by his pessimistic meta-induction. The above argument is then what Lyons rightly calls a "meta-*modus tollens*". It does not help to resort to restrictive explanatory criteria to challenge premise 2 since explanatory successful theories have also been falsified. As a consequence, the no miracle argument is debunked.

According to my version of scientific realism, we have more reasons to believe in the reality of selected theoretical entities and the truth of propositions about them not on the basis of explanationist arguments but on inductive empirical evidence of a causal link between some theoretical entities and observations or measurements. In short, these entities must have been detected. Such selective moderate scientific realism is compatible with fallibilism: we never can be certain that our beliefs about detected entities will not be falsified in the future. However, although I cannot possibly examine all the examples in Laudan's list here, I submit that no argument consistent with the OCIM conditions was (and still is) available to justify belief in any of the entities in Laudan's list. Crucially, the existence of a causal link between observations and these entities (with the properties attributed to them by earlier theories) had not been empirically and inductively confirmed.

Another objection can be levelled against Laudan. He contends that his argument is *naturalist*. It is supposed to have as much strength as a scientific argument grounded on observations. This being so, Laudan is expected to explain why the confirmation of his historical claim (premise 2) is superior to the inductive confirmation of central theoretical claims made by, say, quantum mechanics and special relativity, since these theories have been amply confirmed by very accurate observations whose quantity and variety is impressive. As Michael Levin points out:

> If one is unwilling to trust the inference from data to current scientific theory, how does one come to trust the far more tenuous inference from a few case histories in science's past to conclusions about the future of science as a whole? (Levin 1979, p. 421).[22]

A genuinely scientific theory must clearly delineate the domain of observations that could confirm or falsify it. But Laudan's historical claim does not satisfy this criterion of scientificity. It is hard to identify the kind of historical data that would falsify Laudan's claim (premise 2). To convince ourselves of this, just look at the following formulation of his historical thesis:

---

[22] See also Ghins (1992, p. 255 and 2005, p. 103).

For any theory accepted today by the scientific community, it is likely that in the future it will be recognised that its central explanatory terms are devoid of referents.

Laudan does not of course contend that a theory that has resisted attempts to falsify it for *n* centuries would never be falsified. Let us assume that a theory has been considered true for 3000 years or more, a possibility that can certainly not be ruled out *a priori.* Would this mean that Laudan's thesis has been falsified? No, because this situation could be an exception and the theory in question could nevertheless end up being falsified later on.

Obviously, I do not deny that belief in the existence of viruses, electrons etc. could one day be recognised to be false, since I am a fallibilist. But I maintain that we have better empirically grounded arguments to believe in their existence than not, and also that, given the available evidence, it is unlikely that strong arguments will challenge these beliefs in the future.

3.7 The Stability of Reference Through Theory Change[23]

An important question remains to be examined by the realist: is it more reasonable to believe in the stability of the *reference* of certain theoretical *terms* – especially *natural kind terms*, such as "molecule", "ether", "electron" ... – through the vicissitudes of the evolution of science in the course of history? And under what conditions? When quantum mechanics speaks of molecules, does it refer to the same entities as Perrin's atoms? Does the word "gene" introduced by Wilhelm Johannsen already refer to the gene theorised by contemporary molecular biology? Does the term "dephlogisticated air" used by Priestley denote what Lavoisier called (and that we still call) "oxygen"? When Plücker used the expression "cause of magnetic light rays", was he referring to what J. J. Thomson called "negatively charged particles"? Did the expression "luminiferous ether" designate the same thing as "electromagnetic field"? At first sight, the answer is no. Aren't atoms, genes, oxygen, electrons and the electromagnetic field of our current theories endowed with properties very different from those attributed by earlier theories, and sometimes even incompatible with them?

But let us not move too hastily. Before deciding on this question – which is certainly a philosophical one – we can take advantage of some central distinctions in the philosophy of language. A first distinction, obvious but which must always be kept in mind, is that between the *terms* of language – *words* – and the *entities* they denote or aim to denote – *things*. An expression or term can disappear and be replaced by another that refers to the same thing.

[23] I am particularly grateful to Mario Alai and Diego Marconi for their very valuable remarks on this section.

Another important distinction, dating back to Gottlob Frege in 1892, is the distinction between *sense* and *meaning* (*Sinn*) and *reference* (*Bedeutung*).[24] The expressions "morning star" and "evening star" have different meanings, but refer to the same entity, namely the planet Venus. These two expressions are different *ways* of describing the same planet, or, according to Frege, of referring to it. I will subscribe here to Frege's distinction between meaning and reference without trying to respond to the objections that have been addressed to it by, among others, Bertrand Russell (1905).

The terms "atom", "gene" etc. have seen their *sense* or *meaning* shift as new scientific discoveries have accumulated. Indeed, the *descriptions* of the entities to which these terms are supposed to refer have been cut down or/and enriched with terms referring to new properties, and thus the *way* of referring to the entities they denote (insofar as they denote the same entity) has changed, just as "morning star" and "evening star" do not refer to the planet Venus in the same manner. Thus, the "gene" – originally described as "the cause of transmission of hereditary characteristics" – is now also described as "a segment of DNA (deoxyribonucleic acid) molecule". These changes in the *meaning* of the same terms or expressions have led philosophers such as Paul Feyerabend (1958) and Thomas Kuhn (1970) to claim that successive theories are "incommensurable", that is to say, literally "without common measure".

To refute the incommensurability objection, it must be shown that the *reference* of certain terms can remain *stable* through changes of *meaning*. Isn't an essential condition for comparing theories to assure ourselves that they talk about the same things (at least some of them), even if competing theories may attribute different properties to them? In order to evaluate the respective merits of theories, the denotations of the terms they use must be unambiguously fixed. Otherwise, we would be unable to decide whether they refer to the same things or not. Without referential continuity, scientists debating the merits of competing theories engage in dialogues of the deaf, as Thomas Kuhn (1970, p. 109) famously remarked.[25]

In discussing the properties that should, or should not, be attributed to the electron, for example, what we are talking about must have been previously identified without ambiguity. If the terms that were supposed to refer to certain entities in past theories have no denotation according to present-day theories, scientific progress consists in an accumulation at the level of empirical data only. It then becomes difficult to defend that science also progresses by improving our knowledge of already detected entities and by detecting new ones. It is indeed such a notion of

[24]For an English translation of *Sinn und Bedeutung* see Frege (1984).

[25]However, in the *Appendix* to *The Structure of Scientific Revolutions*, Kuhn says that incommensurability does not imply incomparability. According to him, we can ponder the merits of competing theories by, among other things, looking at the number of problems they solve and the precision with which they are solved (Kuhn 1970, pp. 205–206). Be it as it may, there is no good reason for Kuhn to believe in the (even approximate) truth of statements about genes, electrons etc. since, according to him, history teaches us that there is no reason to believe in the stability of the reference of the terms "genes", "electrons" etc.

progress limited to the level of unaided observations that is advocated by antirealist philosophers, such as Bas van Fraassen.

In order to be in a position to maintain that science is making progress, not only at the level of observations but also in the knowledge of entities to which we cannot have direct observational access, scientific realists should not exempt themselves from proposing a convincing philosophical conception of reference for natural kind terms. They will then be able to rely on this conception to defend that we have good reasons to believe that detected entities exist, while acknowledging that we may be wrong with regard to some of the properties attributed to them. In what follows, I will briefly present and defend the causal descriptivist conception of reference for natural kind terms that I favour.

All philosophical conceptions of reference take proper names as their starting point. A proper name, such as "Albert II" or "Neptune", unambiguously refers to a particular entity. Traditionally, two main philosophical conceptions of reference have been proposed: *descriptivist conceptions* and *causal conceptions*. According to the first, introduced by Gottlob Frege (1892) and Bertrand Russell (1911), the referent is fixed by the *description* of the denoted entity, which consists in mentioning some of the properties it possesses. Such a description is supposed to identify a unique referent. A proper name, such as "Albert II", can be considered a prime example of a word that denotes a particular entity, in this case a single individual person, described as "the sixth king of the Belgians". If no entity has the properties referred to in the description of the name, i.e. no entity satisfies the description, the name has no referent: it does not denote anything at all. This is the case for the names "Pegasus" or "Ulysses", unless it is specified that we are referring to an entity that is merely imagined in some stories. But such cases are not relevant to the issue of scientific realism.

However, by ostension, proper names can also be made to denote a referent without it being necessary to provide a description, or even when the description offered is false. When a baby is given the name "John", there is no need to provide a description if it is present and shown. This is why Saul Kripke (1980) has defended a *causal conception* of reference (anticipated by Ruth Barcan Marcus). According to this causal conception, the referent is fixed by an "act of baptism", by which the person saying the name has the intention to denote a given entity by this name. This act of naming establishes what the name denotes or aims to denote. The name (and its referent) used by the baptizer is transmitted to other speakers through relations between them, i.e. by interactions, which are causal chains. Thus, the same name is used by other speakers who have the same *intention* to denote the same referent, that is, to use the name in the *same way* as those from whom they learned it by interacting causally with them. Typically, a speaker hears the word spoken by the baptizer who uses it for the first time in a certain context by pointing to a particular entity. For example, if I decide to call my pet aardvark "Napoleon", this word – when I use it – refers to my aardvark, not to the French general (even if I learned this name by reading a history book). (Kripke 1980, p. 96) And this same word can be used by other people when they have the same intention to refer to my aardvark. This is why Kripke's theory should be called "causal-intentional".[26]

But, according to Kripke, reference can *also* be fixed by description. He takes the example of the act of baptism of the planet that is the cause of the perturbations of the orbit of Uranus (Kripke 1980, p. 79, note 33 and p. 96, note 42). This planet was named "Neptune" by Le Verrier before it was observed in telescopes. Here, Neptune is described as the *cause* of perturbations in planetary trajectories. This way of reference fixing by description preserves a core element of the causal conception since it is performed by singling out a particular entity and baptizing it "Neptune". The word "Neptune" is then passed on by causal chains to speakers when they intend to refer to the same thing that was initially called "Neptune". (It is essential not to confuse the causal relations between speakers who pass on a name – and its referent – on the one hand, and the causal relations by which entities such as planets cause the appearances observed in telescopes).

Importantly, according to the descriptivist conception, the referent of a name is fixed by a description that contains predicative terms. These terms must also have a referent, namely a property or a relation, otherwise the name would lack one. The referent of a predicate can be fixed by description with the help of other predicative terms. To avoid the pitfall of an unstoppable regress, some predicates must have their referent fixed by *ostension*. This is why successful reference for predicates is ultimately based on the ostension of observed properties, monadic or relational.

As will be shown below, I will embrace a specific conception of reference, which is *causal descriptivist*, for some natural kind terms. I will refrain here from entering into a detailed examination of the difficulties encountered by both the causal and the descriptivist conceptions.[27] For Kripke, the referent of a proper name, such as "Neptune", may be fixed by the description of its referent in some contexts, as we have just seen. It is true that the word "cause" is mentioned in the description of the referent. But it is precisely because the word "cause" *is part of the description* that we can consider this way of identifying the referent as fulfilling the descriptivist conception. The description is quite precise since it does not just say that Neptune is "whatever" is causing the disturbances, which would be a purely *external* characterisation of the referent. The description also mentions an *internal* property of the intended referent,[28] namely that it revolves around the sun. According to the proposed description, if it had turned out that the cause of the disturbances was something else, the word "Neptune" would have been lacking denotation.

However, instead of being interested in the reference of proper names, scientific realists focus on the reference of terms such as "planet", "atom", "electron", "gene" that refer to natural species or *natural kinds*. In Chap. 5, I will defend a reductive and anti-essentialist ontological conception of natural kinds according to which a natural

[26] I owe this suggestion to Diego Marconi.

[27] For an illuminating review of these difficulties, see Reimer and Michaelson (2019).

[28] The property of revolving around the sun is implicitly mentioned by using the natural kind term "planet".

kind is nothing else than an ensemble of specific entities, that is, an ensemble or bundle of specific associated *properties*. But here, I am mainly interested in questions related to the stability of reference of natural kind terms.

Of course, scientific realists are more interested in natural kinds which are not directly observed and whose referents cannot be fixed by ostension. These referents must therefore be identified by description, provided the predicates mentioned in the description designate (directly or indirectly) observable properties. In the next section, I will attempt to show that the causal descriptive conception of reference for natural kind terms can help in solving the problem of the stability of reference through theory change by relying on Robert Nola's work on the case of the electron (2008).

3.8 The Stability of Reference in the Electron Case

By applying the causal descriptivist conception of reference to natural kind terms, Nola claims that the *cause* of what Plücker first observed in 1857 in electric discharge tubes is the same as what we today call "electrons". Consider a glass tube filled with an extremely rarefied gas and equipped with an anode (positive pole) and a cathode (negative pole). Inside what we call today an "electric discharge tube", a high electrical potential is set up between a (positive) anode and a (negative) point-like metallic cathode. Under precise and repeatable conditions Plücker observed – among other phenomena – luminous rays emanating from the cathode and greenish glows on the walls of the tube at the end opposite to the cathode. There must be a "something" (whatever it is intrinsically) which produces this observed effect.

This "something" can be initially identified by means of the following description: *the cause of the rays and the greenish glow (and other phenomena), under specified experimental conditions shown by ostension*. This causal description can be expanded and detailed by mentioning observational predicates referring to the properties of the experimental set-up.[29]

Plücker called the observed rays "rays of *magnetic* light" because the locations of luminous phenomena vary with the positions of a magnet placed near the tube, whereas ordinary light is not sensitive to magnetic fields. The expression "ray of magnetic light", irrespective of its possible meaning, is a natural kind term that refers to a class of things satisfying the causal description above which refers to *extrinsic* properties of the cause only, excluding any mention of its *intrinsic* properties. As Robert Nola aptly puts it:

[29] I assume that the term "cause" is an observational term, although the cause here is not directly observable. Causal links can be empirically established by the Millian methods of agreement, difference, concurrent variation etc. as we have seen.

> (...) Plücker's investigations provided sufficient information to yield identifying conditions for an entity of some kind which caused what Plücker observed, even though no intrinsic property of the kind was known of the entity but only its extrinsic properties in specified experimental set-ups. (Nola 2008, pp. 174–175).

Notice the similarity between the reference fixings of the natural kind term "ray of magnetic light" and of the proper name "Neptune" by Kripke (except that his description mentioned an internal property of the cause, namely being a planet). Notice also that there could have been several causes of the phenomena occurring in discharge tubes. Then, the above description would have referred to a *genus* with several *species*. But in this case it turned out that the researched cause is unique since there is only one kind of electron (as far as we know today).

One must refrain from adding to the description hypothetical *explanations* mentioning *internal* properties which may fail to actually be instantiated. If we do, the description offered above could not be strictly satisfied by any referent. Nola recommends restricting descriptions to "what we can observe or experimentally determine." And he adds "So restricted, the denotation fixing description is less likely to pick out nothing, or to carry excess explanatory baggage which takes us in the direction of incommensurability." (Nola 2008, p. 180).

Later on, the description based on Plücker's observations was enriched by mentioning newly discovered phenomena. By placing a Maltese cross inside a discharge tube equipped with a point-like cathode, Hittorf observed that a shadow of the cross appeared on the wall of the tube opposite to the cathode. This fact provided indirect evidence that the "something" emitted by the cathode travels in straight lines. Hittorf used the natural kind expression "glow rays" to refer to the identified cause. Then, in 1876, Goldstein, by using spatially extended cathodes managed to show that what is shot off by the cathode is emitted perpendicular to the cathode plate, unlike ordinary light which is emitted in all directions. He introduced the expression "cathode rays" to refer to the cause of phenomena. Despite their different meanings there is no doubt that the referring expressions "glow rays" and "cathode rays" – which function as natural kind terms – all refer to the same things, albeit in different ways as Frege would say.

We know today that the cathode emits particles with specific values of (negative) charge and mass, that is, electrons which are responsible for the production of the observed effects. J.J. Thomson is credited with the *discovery* of the electron (in 1897) because he showed that the cause of the luminous phenomena in discharge tubes are particles with *intrinsic* properties of mass and charge. Most importantly, Thomson also succeeded in measuring the mass to charge ratio *m/e*. But *these intrinsic properties are those possessed by what Plücker and others had previously described and identified as the cause of the specific effects observed under the same experimental conditions (or at any rate, under very similar experimental conditions).*

On the basis of experiments, Thomson provided a description which mention *intrinsic* properties identifying a referent which is identical to the one identified by means of external properties only. The cause posited by Plücker and others was not an indeterminate X since they gave detailed *descriptions* of the conditions in which the effects were observed, and these observations could be reproduced by many experimenters.[30] To say that entities of some kind cause specific phenomena is tantamount to providing a description, admittedly extrinsic, of this kind of entities. If there are entities which satisfy such descriptions, then the expression "cause of luminous phenomena in discharge tubes" has a unique referent variously denoted by terms such as "glow rays", "cathode rays", "electrons" etc. The stability of the referent is not endangered by the linguistic variations of the kind terms used and their possible explanatory connotations. Moreover, these connotations are not present in extrinsic causal descriptions.

The *meanings* of referring terms and successive descriptions can differ, as we saw. Yet this does not prevent mutual understanding among scientists. The terms in the causal descriptions were perfectly well understood by all scientists involved in the debates on the intrinsic nature of the cause of phenomena observed in discharge tubes. In the absence of such a common identification of the same referent, it would have been impossible to discover its intrinsic properties by new experiments and observations.

Various descriptions have historically been proposed to designate and identify the cause of luminous phenomena in discharge tubes. But the shorthand natural kind terms corresponding to these descriptions often mentioned possible intrinsic properties of the cause. These terms are obviously theory laden. Crookes mentions supposed intrinsic properties of the cause by using the terms "molecular torrent", "charged matter", "radiating matter" (Nola 2008, p. 185). Stoney introduced the term "electron" in 1891 to denote the smallest unit of electric charge. This term survived, partly because it suggests that the cause is electrically charged. J.J. Thomson, who speaks of "negatively charged corpuscles", experimentally determined that the value of the ratio of mass to charge *m/e* of these corpuscles is exceedingly *small*. Thus, these corpuscles could not possibly be ions of atoms or molecules, which, being much more massive, would exhibit a much higher value of the mass to charge ratio.[31] Besides, the terms "charge" and "mass" were no doubt well understood by all scientists interested in what caused luminous phenomena in electric discharge tubes.[32] It then becomes difficult to defend an incommensurability *à la Kuhn* in this context.

[30]Thus, the referent identified by Plücker doesn't fall prey to the critique addressed by Stathis Psillos to purely causal conceptions of reference (Psillos 1999, p. 286).

[31]By measuring the deflection of cathode rays in electric and magnetic fields, and using Newton's second law together with Lorentz' force (Arabatzis 2012, pp. 98–99), Thomson managed to prove that the value of the *m/e* ratio of the charged corpuscles was about a thousand times smaller than the value of this ratio obtained by electrolysis for the hydrogen ion (Whittaker 1973, pp. 362–363).

[32]Yet Arabatzis pointed out to me that charge was then *commonly* understood as a complex epiphenomenon of the interaction between ether and matter, which does not correspond to what we *mean* by "charge" today.

3.9 Extrinsic and Intrinsic Properties

According to Robert Nola, referential stability is ensured by embracing a minimalist position which consists in identifying a natural kind by a description which mentions only extrinsic observable properties of its members. In particular, it is possible to identify a natural kind of entities by mentioning the observed properties of their effects under specified experimental conditions. To be able to identify an unobserved referent and secure ontological continuity, it is sufficient to mention extrinsic observational properties in the descriptions of natural kinds such as electrons, genes etc. In this way, we succeed in establishing a solid basis for the referential continuity of theoretical terms such "glow rays", "cathode rays", "charged matter", "streams of electrons" etc. despite historical variations in their linguistic expressions. The antirealist objections of a Feyerabend or a Kuhn have then been refuted.

By contrast, for those who, such as Stathis Psillos, claim that the identification of unobserved referents cannot be achieved unless their description mentions some of their *intrinsic* properties, the attribution of which is justified by *explanationist* requirements, it immediately follows that the success of reference crucially depends on some theoretical context:

> Since there is no theory-independent way to identify the causal origin [of phenomena], what bears the burden of reference is these causal-explanatory descriptions of the kind-constitutive properties. (Psillos 1999, p. 289).

For Psillos, reference fixing *always* requires the attribution of some *intrinsic* properties: if we do not mention some intrinsic properties of the referent, namely the properties which play a causal-explanatory role, we remain ignorant of what we are talking about. However, as we just saw, the "burden of reference" can be carried without resorting to intrinsic properties, at least when we focus our attention on a certain type of phenomena (such as luminescence) occurring in accurately described experimental devices (such as electric discharge tubes).

If Nola's causal descriptive approach of reference fixing is correct, it provides a welcome alternative to Psillos' causal explanatory proposal which faces several difficulties. First, Nola's account does not involve any explanatory description of the causal role supposedly played by specific *intrinsic* properties and thus does not fall prey to the objections levelled against explanationist strategies. Second, Nola's causal extrinsic descriptivist conception of reference eschews Psillos' recourse to (hopefully stable) "core causal descriptions" (Psillos 1999, p. 295) which mention the *intrinsic* properties supposed to perform the alleged causal role. This is why the stability of reference of Nola's identifying extrinsic descriptions is not threatened by possible radical theory changes in the course of history.

The position defended here is similar to the one advocated by Arabatzis (2006) for whom the description of experimental devices is crucial when it comes to identifying the unobserved referent of what he calls a "concept". (For my part, I prefer to hold on to language *terms*, in order to avoid the mental or psychological connotations that could be associated with the notion of concept.)

For Arabatzis, the electron concept "continues to refer to the same 'thing' as long as its experimental manifestations remain stable or grow more or less cumulatively." (Arabatzis 2012, pp. 162–163). Among the stable manifestations of the electron concept, Arabatzis mentions the ratio *m/e* whose constant value is what he calls the "signature of the electron" (Arabatzis 2006, p. 43). Obviously, this ratio is an intrinsic, not extrinsic, property of electrons. And it is true that varied, independent and concordant measurements provide a good justification for believing that electrons have the properties of charge and mass whose ratio is specific to them. However, Arabatzis (2006, p. 259) claims that the robustness of the reference here does not depend on high-level theories, but on theories that are close to experiment, that is, *low-level theories*. Yet this point is debatable because Thomson's measurements rely on Newtonian mechanics, which could hardly be considered a low-level theory.[33]

This brings us to come to grips with an important difficulty. How is it possible to identify the same entity through very *different* experimental contexts? Nola's minimalist conception of reference secures diachronic referential continuity when we look at the same phenomena (luminous for example) in similar experimental contexts (such as discharge tubes). But it does not guarantee the identification of the same unobserved referent when we are working with significantly different experimental set-ups. How can we ascertain that the cause of some phenomena in a given experimental context is the same as the cause of other phenomena produced by means of other experimental devices? Here, it seems that we can no longer rely on a description which mentions extrinsic properties only, but that we must resort to intrinsic properties, justifiably believed to remain invariant through the various methods of observation or measurement.

At this point, we must rely on my previous vindication of the moderate version of selective epistemological scientific realism, based on the conditions of causality and invariance discussed in the first section of this chapter. If measurements (carried out by relying on previously inductively confirmed causal laws) of the *m/e* ratio give (approximately) the same value, we have good reasons to believe that we are dealing with entities which have this *m/e* value and then that these entities are identical to the ones previously identified by Plücker, Hittorf, Goldstein, J. J. Thomson etc. This is what happened historically. When, instead of a metallic cathode, some radioactive material was used as a source of beta rays in discharge tubes, similar luminous phenomena were observed, and the *m/e* ratio was again very small compared to this ratio for ionised atoms.[34]

Also, quantum mechanics and special relativity attribute the same *m/e* value to an electron at rest. Therefore, an intrinsic property can remain invariant while theories

[33] In a later article, Arabatzis acknowledges that stability of reference may depend in some cases on the permanence of higher-level theories. (Arabatzis 2012, p. 163)

[34] The *m/e* ratio for beta rays is larger because their electrons have a much higher speed, and therefore a larger mass, than those in the cathode rays. Yet its value is sufficiently small to avoid confusion with ions and other particles.

and laws about an entity's behaviour evolve, as was the case when quantum mechanics was introduced. In this sense, the identification of a referent has a robustness with respect to theory changes insofar as some of its intrinsic properties remain unchanged. It is therefore necessary to distinguish the laws that describe the behaviour of an entity, on the one hand, and the laws that are used to obtain values for its intrinsic properties in an experimental context, on the other. Surely, Newton's laws (approximately) describe electron behaviour, but they are also the laws that were used by Thomson – and for good reason – to determine the *m/e* ratio of the electron with the help of a precise experimental device. However, this is not always the case, as can be illustrated by a famous episode in the history of physics.

At the beginning of the twentieth century, physicists strived to discover the law of variation of the *m/e* ratio for the electron as a function of its velocity. To achieve this aim, various experiments were carried out, among others by Kaufmann. For Lorentz and Einstein, in a given reference frame, the electron mass varies in function of its velocity, in conformity with special relativity. However, Kaufmann's measurements were at variance with the predictions of special relativity, and more in agreement with the electron theory proposed by Abraham.[35] The identification of the electron was not problematic since measurements had to do with the deflection of beta rays in electromagnetic fields, and these beta rays could be identified in electric discharge tubes as electrons. Moreover, their *m/e* ratios were still very small compared to that of ionised hydrogen.

These conflicting *theories* attributed different *meanings* to the term "electron" since they contained different laws of variation of the *m/e* ratio with velocity. But the main issue was not about meanings, of course. The debate hinged on what are the correct laws for the behaviour of electrons. The discussions focused on how experimental devices work, how electrons interact with the fields and on possible sources of error. Eventually, thanks to precise and accurate measurements special relativity gained the upper hand. Clearly, this competition between two theories could not have been resolved if electrons had not been unambiguously identified by all participants in the debate in the first place.

But the course of history never stops. Non-relativistic quantum theory and quantum electrodynamics (QED) describe the behaviour of electrons in very different ways. How can we defend ontological continuity when the successive Hamiltonians (expressing global energy by means of momentum and potential energy) vary and the notion of particle even becomes controversial?[36] In fact, these successive Hamiltonians are different ways of referring to the same things, namely electrons, with increasing degrees of accuracy in the description of their behaviour as attested by their experimental manifestations.

---

[35] For a detailed study of this fascinating episode of the history of physics, see Potters (2019).

[36] For example, the number of particles may depend on the reference frame. In this case, number is no longer invariant but is a perspectival notion, analogous to the various profiles of ordinary objects when seen from various locations. This also holds for the m/e ratio. Nevertheless, particles still are numbered and have a charge and a mass, which vary according to known laws.

For sure, the invariance of the values of intrinsic properties gathered by various measurement methods provides further evidence in favour of referential continuity. Indeed, despite the implementation of different theories in a variety of experimental set-ups, the stability of measurement results supports our belief in the ontological continuity of reference through theoretical changes.

Evidently, scientific realists could not rest satisfied with referential continuity regarding entities of which nothing could be said except that they are the "cause of certain phenomena". In science, we seek to know the nature of the causes of phenomena. To reach such objective, we must be able to determine what are the *intrinsic* properties of these causes and how they relate to their effects. Otherwise, we would remain utterly ignorant of the mechanisms underlying the production of the observed effects. In order to achieve knowledge of these mechanisms, I proposed earlier that belief in the reality of detected (indirectly observed or measured by means of reliable instruments) properties must be vindicated by the empirical and inductive confirmation of a causal link between these detected properties and observations. Granted, belief in detected properties depends on inductively confirmed laws and thus is not independent of any theoretical context. Yet if the theoretical context changes and different intrinsic properties come to be attributed, it will be about referents identified by extrinsic properties and perhaps by some stable intrinsic properties also.

For molecules, the intrinsic properties that play a causal role are the properties of mass and motion, as shown by Perrin. For the electron, these properties are values of charge and mass. As for the ether, Psillos plausibly argues that some, but not all, causal properties attributed to the ether are also properties of electromagnetic waves (Psillos 1999, p. 296). Yet it would be hard to maintain that the luminiferous ether (the medium of propagation of light waves) *as intrinsically described* by nineteenth century physicists exists.[37] This is also true for the phlogiston, the crystalline spheres, the physiological humours and the other entities mentioned in Laudan's list. Laudan is certainly right in claiming that we have no reason to believe in the existence of these intrinsically described entities.

To counter Laudan's objection, I have followed Robert Nola in proposing to resort to a causal descriptivist conception of reference for natural kind terms. In some cases, we can rely on a description that is limited to extrinsic properties only. But when experimental contexts change, there is no other way to rescue stability of reference than resorting to some intrinsic properties provided that belief in their existence is justified in accordance with the OCIM requirements. Given this, it seems to me that it is possible (a fleshed out discussion is not in order here) to maintain that the cause of the combustion phenomena observed by Priestley and Lavoisier – namely, oxygen – is the referent of what Priestley referred to as "dephlogisticated air", as Kitcher and Boantza also claim[38]; that the molecules which are the cause of the phenomena observed by Perrin are also those which are composed of atoms,

[37] On this point, I agree with Diego Marconi (2009, p. 134) and many others...

[38] See Kitcher (1993, pp. 97–100), Boantza (2007, p. 512) and Ghins (2013b, p. 26).

subsequently shown to be made up of protons, neutrons and electrons; that the genes responsible for the appearance (or disappearance) of certain apparent (phenotypic) characteristics of the parents in their offspring are also composed of DNA molecules; that "luminiferous ether" refers to the cause of luminous phenomena, and that this ether is what we call today "electromagnetic field" etc.[39]

It may be objected that it would be difficult to accept a referential continuity with respect to, for instance, "crystalline spheres" and "gravitational forces" when referring to the cause of planetary motions. Nevertheless, this did not prevent scientists from understanding each other reasonably well when they argued about the nature of the cause of the stability of planetary orbits, whatever the kind terms "crystalline spheres", "angels", "magnetic forces" etc. that were historically used to refer to the cause of such stability. All understood that they were talking about something minimally described as the cause of the apparent regular permanency of planetary motions, irrespective of the various intrinsic properties dissenting participants in the discussions were attributing – justifiably or not – to it.

In line with what has been said above, realists acknowledge that the referents of terms like "gene", "electron" etc. do not satisfy the various descriptions that have been proposed for them in the course of history. These referents do not possess all the *internal* properties mentioned in the descriptions accepted by scientists at a given time. Descriptions of a referent by means of its *intrinsic properties* have sometimes varied considerably throughout the history of science. Scientists have very often believed (for explanatory reasons) in the existence of some of the internal properties mentioned in the description of a referent even when this belief was not supported by arguments complying with the four OCIM conditions. What is more, new properties of the cause of certain phenomena may be discovered later. Some kind (*genus*) may have several sub-kinds (*species*).[40] It is not excluded that we will one day discover that there are two kinds of electrons.

Be it as it may, the identification of "what we are talking about" is first ensured by a description of the *extrinsic* property of being the cause of clearly characterised phenomena. Moreover, changes in the *meanings* of expressions denoting the same referent do not imply that proponents of different theories about *intrinsic properties* do not understand each other due to semantic incommensurability. I have abundantly insisted that the identification of an unobserved referent must be first achieved by relying on immediate experience, such as direct observation of light phenomena in

---

[39] An anonymous referee objected that if the ether exists, then the god Jupiter also exists, since he is the cause of lightning. Granted, the description "the cause of lightning" can be given an arbitrary name, and this name could be "Jupiter". But if "Jupiter" denotes something having particular intrinsic properties, as was the case in Greek mythology, there was no cogent justification to believe in the reality of such referent. There is no way to empirically ascertain, by Millian methods, that Jupiter, with the intrinsic (unobserved, though observable) properties attributed to him, is the cause of lightning. Such Jupiter is not available for reproducing and manipulating lightnings. This also refutes Quine's assertion that physical entities such as electrons are "comparable, epistemologically, to the gods of Homer" (Quine 1951, p. 41).

[40] See the discussion of the chlore isotopes in Chap. 5.

discharge tubes. In such cases, scientists can perfectly well understand each other and dialogue among themselves, even if they disagree about the intrinsic properties that should be attributed to these entities. What is more, intrinsic properties the belief in which is justified on the basis of the four OCIM conditions manifest remarkable stability through theory change. The mass/charge ratio for the electron is a case in point.

3.10 Conclusion

In this chapter I have tried to defend a moderate version of scientific epistemological realism within the framework of the synthetic approach to theories. This realism concerns both directly observable entities and those that are indirectly observable, i.e. detectable. To justify belief in the existence of entities with detectable properties, we must summon arguments similar to those mobilised in the context of our everyday experience, by deploying a bottom-up approach that takes the immediately observed properties of ordinary things as a starting point. In the same way that we assure ourselves of the existence of observed properties of ordinary objects by repeated and varied observations, we convince ourselves of the existence of detected properties by relying on various independent measurement methods that deliver concordant results. At the same time, empirically proven causal relationships between the observations or measurements and the detected properties of entities of a certain kind ensure that we have succeeded in establishing a causal link with the latter, without invoking a problematic inference to the best explanation. In short, I proposed four normative OCIM conditions that must be satisfied by an argumentation aiming at justifying belief in detected properties. With this proposal, I hope to have shown that my inductivist defence of epistemological scientific realism does not fall foul of the antirealist objections of the underdetermination of theories by observations.

By way of illustration (not demonstration) of the ability of the bottom-up approach to support the realist position, I have shown that the four OCIM epistemic requirements are satisfied by the argument for belief in the existence of Neptune and Perrin's argument for the reality of molecules. I have also responded to Laudan's pessimistic meta-induction by relying on a causal descriptivist conception of reference, as proposed by Robert Nola. Thanks to this causal descriptivist conception for kind terms, it is possible to argue in favour of the ontological continuity of some referents, such as electrons, across new empirical evidence and changes in theory despite the variety in meaning of the linguistic terms used to designate them. Besides, belief in intrinsic properties which is inductively and causally supported shows impressive stability through theory changes. All this allows us to counter the antirealist objection based on the alleged incommensurability of meanings brandished by, among others, Kuhn and Feyerabend.

The version of epistemological scientific realism I proposed is moderate since it is selective and also parsimonious because it does not vindicate belief in undetected as well as *a fortiori* non-detectable entities. Moreover, moderate realists do not claim to have developed a way of arguing that precludes error, since they have to accept revision of their beliefs when confronted with new and unexpected empirical data. However, in the absence of such data, it remains reasonable to believe in the reality of some entities that are not directly observable, provided – as I have abundantly emphasized – that they enjoy an inductively verified causal link with direct observations or measurements (that is, they have been detected) and that various measuring methods give concordant evidence for their existence. These conditions are very strict. This is why such beliefs are always partial and limited.

Chapter 4
The Categorical Conceptions of Laws

Abstract This chapter is devoted to a presentation and critique of the categorical conceptions of laws. For categorical monists, only categorical properties exist; powers and all other dispositional properties can be reduced to categorical properties, namely, their categorical bases. Before clarifying the all-important distinction between categorical and dispositional properties, we must be clear on what a property is. I argue in favour of construing properties as universals. After that, I briefly address the issue of the distinction between categorical and dispositional properties. I then proceed to discuss the regularity account of laws, which has been initiated by David Hume and developed by Mill, Ramsey and Lewis, and I elaborate on the reasons why the various versions of regularism are unsatisfactory. In the last part of this chapter, I examine how Armstrong, Dretske and Tooley strive to overcome the difficulties encountered by the regularists by introducing a contingent relation of necessitation between properties, while remaining faithful to categoricalism and Humean contingency. I conclude that categorical accounts of laws, whether in their regularist or necessitarian versions, face insuperable difficulties.

For Nancy Cartwright (1983), Bas van Fraassen (1989) and Ronald Giere (1999), the concept of law, in addition to being problematic, is simply useless. For them, not only is it impossible to provide a satisfactory philosophical characterisation of the concept of law, but it does not help to understand what science is. Indeed, if one upholds a semantic conception of theories – as Cartwright, van Fraassen and Giere do – laws give way to models, which then attract the full attention of philosophers of science. In compliance with the synthetic conception of theories presented in the first chapter, I maintain on the contrary that abandoning the notion of law would lead to an impoverished conception of scientific theories and would impair our understanding of what science is.

For a universal proposition to deserve to be called a "scientific law", it must belong to a scientific theory, and we must also have good reason to believe that such proposition is (at least approximately) true on the basis of a sound argument that satisfies the four OCIM conditions stated in the previous chapter. This criterion of legality or nomicity is *epistemic*. It allows us to expel from the paradise of nomicity

M. Ghins, *Scientific Realism and Laws of Nature: A Metaphysics of Causal Powers*, Synthese Library 483, https://doi.org/10.1007/978-3-031-54227-5_4

not only isolated propositions, such as the too famous "All crows are black",[1] but also accidental generalisations like "All the coins in my pocket are euros",[2] since these propositions are not part and parcel of any scientific theory. Being a proposition, a scientific law is not a reality in the world. I therefore defend a *propositional* (also called "enunciative" or "syntactic") philosophical conception of scientific laws, distinct from *ontological* conceptions, according to which laws are real entities. However, factual regularities that make scientific laws true are real.

Laws that are approximately true for some systems are not true of others. According to the objectifying approach to science, a gas is an entity that instantiates a system of properties (volume, pressure, temperature). This system or structure of properties makes the Boyle-Charles formula true, as I showed in Chap. 1. Similarly, a simple pendulum is an entity identified by means of immediately observable properties. The law of isochronism of small amplitudes, attributed to Galileo, does not describe the behaviour of systems that are not simple pendulums.

A scientific law asserts that particular entities exemplify relations between properties, either statically, such as the ideal gas law, or successively, such as Newton's second law. In accordance with the synthetic conception of theories, laws describe structures of properties or temporal successions of structures of properties. As spelled out in the first chapter, these structures are the ideal models that exactly, or nearly so, satisfy the laws. These models can also represent, with various degrees of precision, real structures of properties instantiated in particular entities. Scientific laws describe, more or less accurately, the structures of properties instantiated in these real entities.

Unlike most authors, I distinguish *scientific* laws from laws of *nature*. *Scientific* laws acquire the status of laws *of nature* if their nomicity is based on a *metaphysics* of nature. This means that the truth of the statement "L is a law of nature" (where "L" denotes a scientific law) rests on the existence of properties which are metaphysical, since they cannot in any way be directly observed or detected. Given that modal properties (possibility, necessity) are not accessible to sensory perception, strict empiricists deem it unjustified to attribute some kind of necessity to laws. According to strict empiricists, scientific laws are only contingently true. But, as I will try to show in the last chapter, we have good reasons – based on experience – to believe that there are metaphysical entities, namely powers, which ground the *necessity* – in a sense that will have to be made precise – of scientific laws. Scientific laws are made true by regularities, but their lawful character or nomicity is grounded on the existence of powers. The terms "scientific law" and "law of nature" have the same extension (they denote the same propositions) but their meanings are different: the latter expression involves necessity, whereas the first one does not.

---

[1] Even if isolated universally true propositions are not *scientific* laws in my view, this does not mean that they are irrelevant to science. They can contain useful information and may be later included in scientific theories.

[2] This is of course Nelson Goodman's well-known example, in a slightly modified form (Goodman 1955, p. 42).

In this chapter, I will present and criticise the so-called "categorical" conceptions of laws. For categorical monists (or categorists, for short), only categorical properties exist; powers and all other dispositional properties can be reduced to categorical properties, which are called their "categorical bases".

Before attempting to clarify the all-important distinction between categorical and dispositional properties, we must be clear about what a property is. This is the issue I will tackle first. After arguing in favour of my preference for a construal of properties as universals, I will address the question of how categorical and dispositional properties can be distinguished. I will then proceed to discuss the regularity accounts of laws, which were initiated by David Hume and developed by Mill, Ramsey and Lewis. In the last part of this chapter, I will examine how Armstrong, Dretske and Tooley strive to overcome the difficulties encountered by the regularists by introducing a relation of necessitation between properties, while remaining faithful to categoricalism and Humean contingency. I will conclude by showing that categoricalism, whether in its regularist or necessitarian version, does not offer a convincing solution to the problems of grounding the nomicity of laws and their relation to occuring regularities.

4.1 Particulars and Universals

So far, I spoke of entities and properties, without saying much about what they are. Yet we are here confronted with a fundamental problem which, from the very beginning of philosophy, has generated passionate controversies. They are still very much alive today! But, since this is not the place to discuss this issue at length, I will limit myself to a succinct argumentation in favour of the view that particular entities are sets (and sometimes structures) of instantiated universal properties at some spatiotemporal location. In what follows, I will mostly be inspired by David Armstrong's work.

Most empiricist philosophers subscribe to nominalism,[3] according to which there exist particular entities only. Nominalists thus profess antirealism with respect to universal properties. On the contrary, for the opponents to nominalism,[4] some predicates denote universal properties, that is, real universals. A *universal* is repeatable, while remaining perfectly identical. It can be instantiated in a variety of particulars while remaining the same property. A universal is not a particular, since it can have an indefinite number of instantiations while remaining identical. This is also true for the specific values of determinable properties such as volume or

[3]There are notable exceptions, however. Empiricists such as Hans Reichenbach and David Lewis were opposed to nominalism.

[4]Armstrong gives a clear presentation of the various forms of nominalism in his wonderful book on universals. (Armstrong 1989)

pressure. A determinate precise value of volume for example can in principle be instantiated an indefinite number of times.

Notice that not all predicates refer to real properties. For example, artificial or constructed predicates, such as "being blue or spherical", do not refer to any property, as we shall see below. We must then distinguish, as David Lewis (1986b) and others do, between *abundant* and *sparse* predicates. Only sparse predicates denote real properties; they are also called "natural properties". Typically, the predicates mentioned in scientific theories which refer to properties that are observable in principle, whether directly or indirectly, are natural predicates.[5] Obviously, only natural properties will interest us here.

Various arguments have been adduced in favour of the existence of universal properties. First of all, it has been claimed that if one accepts the existence of universals, one can *explain*[6] the observed resemblance between entities. Hard entities resemble each other because they all instantiate the *same* universal: the property of being hard or hardness. Postulating the existence of universals would be the only, and therefore the best, explanation of observed similarities. As Stathis Psillos graphically says, universals are supposed to be *resemblance enforcers* (Psillos 2014b, p. 12): they compel individual entities that instantiate the same universal to manifest a similar appearance in some respect. Instantiating the property of hardness "forces" a particular entity to manifest this property. As for the differences between particulars, they would be explained by the instantiation of different sets of universals.

However, as we saw, we have every reason to be sceptical of explanationist strategies, especially when the clarity of the explanation offered is questionable, as is the case here. What is the link between the universal and the particular entity in which it is instantiated? What is the nature of the instantiation relation? The answer to this question requires clarifying what a particular is, without getting embroiled into intractable metaphysical problems on the nature of the relationship[7] between a particular entity and its properties.

How shall we then construe a particular entity or thing? If we mentally strip a particular of all its properties, we get the notion of a pure substratum or *bare particular*, although it is quite impossible to imagine or to have epistemic access to a thing having no property. This is why we have no reason to believe in the existence of bare (or thin) particulars. I will thus hold that a real particular is always clothed or thick: it possesses some properties. It is not an underlying something of which nothing could be said. A particular cannot be a pure substratum, like Aristotle's prime matter. No, a particular has properties: those which are instantiated in it.

Rather than going down the road of explanation, let us focus on experience. If the proposition <this observed entity is hard> is true it has a *truthmaker*, which is a fact

[5]"Having a mass", "having a charge" etc. are natural predicates.

[6]For an explanationist defence of universals, see Armstrong (1989, p. 103).

[7]If instantiation is taken to be a relation between a property and a particular, this relation is instantiated when the particular *a* has the property F. What is then the relation between the instantiation relation and its relata *a* and F? We immediately see that we are caught up in an endless regression, as Bradley (1963) pointed out.

that corresponds to it in accordance with the truthmaker principle.[8] The truth or falsity of the statement "F*a*" hinges on the instantiation or not of a monadic property[9] F in *a*, which constitutes the *fact* (or *state of affairs*) F*a*. F*a* is a thick particular, a fact or state of affairs. The thin particular is the particular abstractly considered as being stripped of all its properties (Armstrong 1989, pp. 94–96). The thin particular is a logical fiction; only thick particulars exist, which are nothing else than instantiations of properties. There are no such things as property-free entities, or anyhow we do not have any reason to believe in their existence, since in experience we can only be acquainted with thick particulars.[10] To be epistemically accessible, a particular must instantiate at least one observable property.

For a relational statement "*a*R*b*" (e.g. "*a* is greater than *b*"), the question of its truth (or falsity) is a matter of instantiating (or not) a relational property R (such as the dyadic property of being greater than), by particulars such *a* and *b*. The statement "*a*R*b*" is satisfied by the fact *a*R*b*, which is none other than its truthmaker. I will not take *a*R*b* as being a particular but a particular fact, since it involves two particulars *a* and *b*, which instantiate monadic properties even if these are not referred to in the statement "*a*R*b*".

It is thus quite reasonable to admit that there are real properties, externally present at a space-time location (which is always extended), even if this conclusion is not strictly imposed by perceptual experience. In fact, this conclusion follows from the view of truth as correspondence – and its consequence, the truthmaker principle – which is mainly supported by the recognition that new observed facts sometimes compel us to revise our beliefs, as I have shown in Chap. 2. But experience of error is not *stricto sensu* an immediate experience, even if it is concomitant to such experience. It is a meta-experience. In this sense, admitting the existence of real external properties is a view which is not only ontological, but also metaphysical. It could be that properties are just sensory impressions or sense data, as Hume, Berkeley, Russell and many other empiricists have claimed. But if we endorse the truthmaker principle, the existence of external properties is the most plausible view. Henceforth, I will assume that there are real universals.[11]

---

[8]See Chap. 2, Sect. 2.

[9]A monadic property is a property that can be attributed to a single entity. A dyadic property (being smaller than, for example) applies to two entities. We also have triadic (being between) and, more generally, polyadic properties.

[10]From now on the term "particular" will refer to thick particulars, unless otherwise specified.

[11]Armstrong (2004, p. 43) acknowledges that the truthmaker principle does not allow us to decide whether properties are universals or tropes, that is, particularised properties. According to the friends of tropes, the charge of *this* electron is distinct from the charge of *that* electron. Realism about tropes is thus a version of nominalism. Since there are entities – such as the elementary particles – that have exactly the same detectable properties, this seems to support the existence of universals, as we saw above.

The theory of universals is far from being immune of difficulties, especially that of characterising the similarity between universals in a satisfactory way. (Armstrong 1989, pp.103–107) Alternatively, and more economically, one could simply accept the external reality of properties, while remaining neutral on whether they are universals or tropes, as Chakravartty (2007, pp. 81–82) advocates.

As for the nature of particulars, an option is to take them to be *bundles* of instantiated properties, as proposed by Bertrand Russell (1948), and much favoured by empiricists. This conception is more economical since any form of possible substratum is eliminated. But it also harmonises well with what we are acquainted with in perceptual experience, since what we see, hear, touch etc. are properties at spatiotemporal locations. Objects such as tables and chairs are perceptually recognized as such on the basis of the properties we observe.

To put it concisely, a (thick) particular is an *instantiation of properties* (which can also form systems or structures) at some space-time location. As a way of speaking, we can say that a thick particular instantiates properties and that properties are instantiated *in* (and not *by*) thick particulars. As Armstrong says:

> *Thick particulars are states of affairs.* The properties of a thing [thick particular] are "contained within it" because they are constituents of this state of affairs. (Armstrong 1989, p. 95).

A particular then is nothing else than a set or a bundle of instantiated properties. Suppose now that a property F would be a constituent of a particular. Logically, we cannot attribute to a set of properties a property that belongs to that set. If we do, we get a false proposition. Consider a particular which contains the properties of being liquid, colourless, odourless and tasteless. Can we say that this *set* of properties is liquid? No. We can certainly assert that this set contains four elements, and also that its elements are observable properties etc. But we cannot attribute to this set of properties the property that it is liquid, or any of the properties it contains. To refer to such an entity, identified by some of its observed properties, we use an abbreviation: the word "water". By this word "water" we do not refer to an added property but to that which is liquid, colourless etc. In short, we refer to the (thick) particular which contains the four properties mentioned above. Surely, water is liquid etc. which is to say that a sample of water is a *particular entity* that has the property of being liquid as one of its constituents. Moreover, we leave open the possibility that this particular may have other properties, such as having the chemical composition H_2O, and also hitherto unknown properties. Should we wish to, we can also use the word "water" as a predicate, as in the sentence "This is water". The predicate term "water" or "being water" then refers to a complex property, namely a property that is the conjunction or sum of the properties of being liquid, odourless, colourless, tasteless etc. But this way of speaking is misleading, as it could give the impression that an additional, complex property has been added to the properties actually instantiated in a particular, which is not the case. Water is usually taken to be a natural kind. (In Chap. 5 I will propose a metaphysics of laws of nature that does not resort to natural kinds.)

When we claim that a particular, such as a water sample, instantiates the property of being liquid, it is tantamount to saying that the properties of being liquid etc. are components of this particular, which can also be called, as Armstrong does, a "state of affairs" or a "fact". Such way of speaking would be rather convoluted, but it is coherent. It would also be at times analytical and even trivial, if liquidity is among the properties used to identify water, but it would certainly not be analytical if it is claimed that water, identified by superficial properties, instantiates the chemical composition H_2O.

Thus, I embrace a realist position according to which there are real properties which are instantiated in or possessed by real particulars in the sense that real properties are constitutive elements of these particulars. And I am inclined, like Armstrong, to believe in the existence of universals. If a particular *a* instantiates the universal F, we have a fact or state of affairs F*a* that makes the statement "F*a*" true. Of course, particulars are observed thanks to their observable constitutive properties. But we can also refer to things without having to refer to some of their properties. We can use a proper name such as "John" or demonstratives such as "this" or "that", which simply point to a spatiotemporal location. This generally does not raise any difficulty for entities identified in perception. However, as I stressed in Chap. 2, a true proposition *does not represent* a fact, in the regimented sense I understand representation. If in the statement "F*a*" the term "F" refers to a specific property, the term "*a*" does *not* refer to a thin particular which would have no property except the pseudo-property of being referred to by the term "*a*". Again, in perceptual experience we only see thick particulars. So when we say that *a* has property F, we abstract from *all other* properties of the particular we are talking about and consider only property F. In such formulation, we use "*a*" as a *name* to designate a thick particular and not a naked particular, which is a vicious abstraction. As David Armstrong puts it:

> The naked particular and the pure universal are vicious abstractions of states of affairs (facts). (Armstrong 1983, p. 84)

In practice, we must be able to identify, in perception, what we are talking about, namely thick particulars, and never thin particulars, of course. If we say: "this diamond is hard" we must have referred, implicitly at least, to properties *other than* hardness, or, minimally, by saying "this" or "that" in an act of ostension to identify the *a* to which we attribute the property F. Otherwise, we would be condemned to remain imprisoned in language and to state only analytical propositions such as "this hard thing is hard". So, the truthmaker of "F*a*", once again, is indeed a thick particular, a state of affairs or a fact F*a*, of which F is a component, as Armstrong claims. Such conception recalls the classical Aristotelian doctrine of the substance (thick particular) with its attributes (properties). In a proposition, some of the attributes instantiated in substances are explicitly referred to, but not all. However, it is the whole substance, with all its attributes or properties, that is the truthmaker of a proposition that refers to one or more of its properties. (Keeping in mind, however, that it is necessary not to confine the notion of substance to ordinary objects, but extend it to fields, processes etc.)

There remains the vexing question of the identification of the real, natural properties, and of the sparse predicates that refer to them, as opposed to the abundant or artificially constructed predicates. Take for example the property of being (all over) hard or sweet. When we observe a thing which is all over hard, we do not observe the complex property of being hard or sweet. We never observe disjunctive properties at the same time and place. Even if we can logically deduce the proposition <this candy is hard or sweet> (proposition B) from the proposition <this candy is hard> (proposition A), this does not oblige us to accept that the property of being hard or sweet exists. Proposition A – like proposition B – is made true by the fact that the candy is hard and certainly not by the pseudo-fact that the candy is hard or sweet. We do not have the power to create new existing entities a *priori* by means of logical operations. Long before Wittgenstein, Plato, Aristotle and many philosophers have warned us against the traps of language.

4.2 Categorical and Dispositional Properties

In this section, I will present the central distinction between categorical and dispositional properties, a distinction that will be further analysed in Chap. 5. I will also take for granted that the properties instantiated in particulars are real universals.

A property which is in principle observable (directly or indirectly) is a categorical property. As Ellis says: "(...) the categorical properties have long been recognized by empiricists as being among the most directly observable of properties." (2010, p. 139) Surely, the existence of unobservable categorical properties cannot be discarded, even if we have no reason to believe that such properties exist. Thus, we are not in a position to assert that a property is categorical if and only if it is observable in principle. Yet if a property is observable in principle, we know that it is categorical. Although relative to us (of course!), observability permits us to classify some properties (those we might have reasons to believe that they exist) as categorical, without contrasting them with dispositional properties.

Nevertheless, reflecting on this contrast will allow us to identify categorical and dispositional properties ontologically, and not only epistemically on the basis of our possibilities of observation.[12] Let us start with a classic example of categorical property, namely sphericity. It does not require much reflection to realize that sphericity is a property that does not direct the particular in which it is instantiated towards the possible acquisition of another property. Categorical properties are static, not dynamic. They are inert and passive properties. By themselves, they do not confer any possibility of action to their bearers: they are *non-modal* properties. For example, the possibility of rolling is not denoted by the predicate "spherical". Such possibility is referred to by the (complex) predicate "having the possibility of rolling". The predicates "spherical" and "having the possibility of rolling" have different meanings without being distinct ways of denoting the same property. They refer to distinct properties. The instantiation of the property of sphericity does not imply that it has the possibility to roll (since it also would have to be hard to have the capacity to do so). The definition of the spherical shape as a three-dimensional set of

[12] Besides, it may appear paradoxical to characterize categorical properties with the help of a dispositional property, such as observability.

points located at equal distance from a centre does not include any indication of a possible behaviour for something like a billiard ball.

Besides geometric properties, structural properties provide more classic examples of categorical properties. A structural property is a property possessed by a system. Such a property consists of the spatial and temporal relationships between the parts of the particular entity that instantiates it. The organisation of atoms in a molecule or the spatial arrangement of molecules in a crystal are structural categorical properties. Since observability implies categoricity it is legitimate to extend the set of categorical properties to *all* observable properties, as categorical monists do, without restricting them to geometrical and structural properties only. On this I concur with categorists, as I will contend in the next chapter.

Let us now turn to *dispositional properties*. A disposition endows the particular who bears it with the possibility of acquiring other, not previously possessed, properties (and losing some). When instantiated, these other properties are the *manifestation* of the disposition. Such manifestation consists in the instantiation of *categorical* properties. Sugar, for example, since it has the intrinsic dispositional property of being soluble, has the real possibility of acquiring, i.e. of manifesting, the categorical property (or more precisely the temporal sequence of categorical properties) of dissolving in water. For dispositionalists, dispositions are irreducible to categorical properties. They are therefore irreducibly modal. In what follows, I will almost exclusively focus on dispositions whose manifestations are described by scientific laws.[13] Such dispositions will be called *natural* or *causal powers*. For example, the dispositional property of charge is a natural causal power to repel charges of the same sign. Powers are identified by the possible manifestations they endow their bearers with.[14]

The same property cannot be both categorical and dispositional. A categorical property in no way satisfies a description that mentions dispositional predicates. It is impossible for the same property to satisfy incompatible descriptions.[15] Therefore, I cannot follow John Heil (2003) and John Carroll (1994) who claim that the same property can be referred to by both categorical and dispositional predicates, depending on how we look at it.

In contrast to categorical properties, dispositions are dynamic, active properties. The predicate "to be soluble" refers to a property that confers on the entity that possesses it the possibility of engaging in some activity, of actualising a certain behaviour, of changing its state. Since the process by which a particular acquires the manifestation corresponding to the disposition consists in a change with respect to some of its properties, it is a causal process that can be described by a causal law. In

---

[13]More accurately, manifestations are described by nomological formulas, as explained before. But, for the sake of simplicity, I will often use the word "law" to refer to manifestations.

[14]I am following Vetter (2015) here. I will come back to this important point in the next chapter.

[15]The expressions "morning star" and "evening star" do not refer to incompatible properties since they refer to temporally indexed properties. They may therefore belong to the same particular: the planet Venus.

this case, a process described by a causal law can be said to be the manifestation of a power instantiated in a particular entity.

But the manifestations of some powers are not described by causal laws. A gas at equilibrium satisfies the Boyle-Charles formula. If the environment does not change, the gas has the power to remain in the same state. It will not spontaneously change state. It will only move to another state if external circumstances trigger this process. It seems to me that the ability of an entity to stay in a specific state, described by a static law, can also be considered a power.

Causes are events. An event is the instantiation of one or more properties at some spatiotemporal location: they are particulars or facts. Thus, instantiated powers can be causes. They play a decisive role because they contribute to the production of manifestations, which are also particular events (or sequence of them). The occurrence of manifestations depends on the existence of powers. In this sense, the term "causal power", frequent in the literature, is justified. Manifestations ontologically depend on powers. Thus, instantiated powers are causes, *albeit* metaphysical since they are unobservable.[16]

Unlike powers, categorical properties are observable properties. Admittedly, observability is a disposition. A spherical object has the possibility of being observed as such under the right circumstances. But it is spherical regardless of whether it is actually observed or not. For a dispositional property such as solubility, the situation is different. We cannot observe, either directly or by detection, the solubility of sugar. Therefore, it is a metaphysical property. Similarly, and perhaps paradoxically, we do not observe that a property is observable. In some cases, we do observe a categorical property, but we do not observe its observability, even though, logically, we can infer its observability from its actual observation. It is the spherical property that we observe, and we infer that it is observable, since possibility follows from reality. If I observe the property F (always instantiated in some particular *a*), then F is observable. Observability is then a second-order property of categorical properties since it is about a first-order property instantiated by a particular. As far as dispositional properties like solubility, malleability etc. are concerned, they are first-order universals. According to the standard logical order hierarchy, particulars and facts are zero-order.[17]

The division between categorical and dispositional properties is the basis of a neat classification of the various philosophical conceptions of laws, which will be examined later. As an appetiser, here is a brief survey of these.

According to *categorical monism*, only categorical properties exist. Categorical monists reject the inclusion of irreducible dispositional properties in their ontology. For them, all dispositions – such as solubility, fragility etc. – are reducible to their categorical bases, that is to say, the molecular structures underlying them. Solubility

[16] As common in the literature, I will henceforth use the expression "causal power" to refer to all powers, including powers to persist in the same state.

[17] For a detailed presentation of this ordering, see Tooley (1977, pp. 678 ff).

"supervenes"[18] on the set of the categorical properties possessed by the molecular components of soluble substances and the liquids in which these substances can dissolve. The laws that describe the dissolution process refer to categorical properties only. Scientific laws are contingent. Thus, categorical monists remain faithful to the contingentism promoted by David Hume and his successors.

Now, categorists are divided into two opposing groups: the *regularists* and the *contingentist necessitarians*. For the regularists, scientific laws are contingently true, i.e. true in some logically possible worlds,[19] including our own, but not in all. Contingentist necessitarians, on the other hand, argue that laws are *necessarily* true in our world and some others, but different laws are necessarily true in some possible worlds. For them, the necessity of a law is not absolute; it is relative to some possible worlds.

Those whom I will call *dispositional necessitarians* are opposed to both versions of categorical monism. For necessitarians like Brian Ellis and Alexander Bird, laws are necessary, i.e. true in all logically possible worlds. According to them, laws are absolutely necessary. On the other hand, for more moderate necessitarians, such as Stephen Mumford and myself, the necessity of a law is relative to the instantiation of appropriate causal powers in our world. What is more, not all contingentist necessitarians share the same ontology. They fall into two categories: dispositional monists and dualists. For dispositional *monists*, such as Stephen Mumford and Alexander Bird, all natural properties are dispositional: they are powers, in the sense that they confer on their bearers the possibility to acquire other properties (or to keep some of them). According to dispositional monists, the so-called categorical properties actually are dispositional properties. In contrast, *dualists*, such as Brian Ellis and myself, include both categorical and dispositional properties in their ontology. Dualists thus adopt a *mixed* ontology of properties.

4.3 The Regularist Conception of Laws

No one will be surprised to hear that philosophers who profess a strict empiricism draw their inspiration from David Hume and argue in favour of the regularist view according to which nomological propositions describe merely contingent observable regularities. Since we never observe the possibility or necessity of any actual event or process, Hume insists that we are not justified in believing that there are real external necessary causal connections between observed events, even if these events always occur one after the other, without exception. Thus, the immersion of sugar in

[18]The relation of supervenience is an asymmetric relation. The property S supervenes on the set of properties B, if the presence of B implies the occurrence of S. But the property S can be realised by different bases B, B′ etc. So, S supervenes on B, but B does not supervene on S.

[19]Henceforth, when I speak of possible worlds, without further specification, I shall have in view the logically possible worlds, that is to say, the worlds in which the logical laws are true.

water is always followed by its dissolution, under the appropriate circumstances. Similarly, lighting a fire is always followed by the instantiation of smoke. But perceptual experience alone does not entitle us to claim that fire will *always* be associated with smoke in the future, nor that fire is *necessarily* accompanied by smoke. We can easily *imagine* that the presence of a fire is not associated with the occurrence of smoke. The necessity that we are inevitably inclined to attribute to a regular connection arises from our irresistible – but subjective – inclination to infer the presence of smoke from that of fire,[20] and vice versa. For Hume, necessity resides in our mind, it is *internal* to us; no question for him to anchoring necessity in an external reality, existing independently of us.

We only observe the constant and repeated conjunctions of distinct events: the presence of a fire is always associated with smoke. Due to our observation of the joint and recurrent occurrence of certain *types*[21] of events, we have become accustomed to inferring that when we observe smoke, a fire is present (and vice versa). We make this inference spontaneously and are naturally driven to do so by force of *habit*. This is why we attribute necessity to such an inference, which we call *causal*, but which is distinct from logical inference.

The necessity we confer to causal inference has its origin in an internal sense of expectation that is purely mental and devoid of any correlate in reality. Human nature is such that we cannot help making causal inferences, since an impression of fire immediately arouses in our minds the idea, the "weakened impression", in Hume's parlance, of smoke. If we light a wood fire, it is inevitable that we expect to observe smoke. But this does not give us sufficient reason to believe in the independent existence of a necessary causal relation between the presence of fire and that of smoke. Sensory experience does not warrant believing that fire *necessarily* causes smoke. Although, for Hume, the necessary connection is part of the very concept or idea we have of causality, this does not warrant that this necessary connection exists in reality. Observation teaches us that the external *reality* of the relation between cause and effect boils down to the mere constant conjunction of certain kinds of events linked together by relations of temporal priority (causes precede effects) and spatial contiguity (cause and effect are in immediate contact).[22]

In his *Treatise of Human Nature* of 1739, Hume proposes two famous definitions of causality. The first, which he calls "philosophical" and which can also be called *objective* or *real*, is based on what can actually be observed, namely constant temporal successions of contiguous and similar events.

---

[20] Although the proposition <All fires produce smoke> is not a scientific law by my criteria of nomicity, Hume's analysis is applicable to scientific laws of temporal succession.

[21] I use here Peirce's classic distinction between *types* and *tokens*. For example, many coins (*tokens*) are worth one euro (*type*).

[22] Notice that causes are distinguished from the effects by an asymmetrical temporal relationship only, without reference to any intrinsic property which would characterise causes as distinct from their effects. For Hume, the distinction between cause and effect is purely external.

> A cause is an object prior and contiguous to another, such that all objects similar to the first are placed in a similar relation of priority and contiguity to objects similar to the second. (Hume 1978, p. 170)

An object, or, in our terminology, an event, is the instantiation of properties in a spatiotemporal region. Thus, the occurrence of an event is a particular fact. The spatiotemporal instantiation of the (complex) property denoted by the predicate "being fire" is the cause of the occurrence of the property denoted by the (complex) predicate "being smoke". The concept of object, unlike that of event, connotes some stability, that is, a permanency in the instantiation of some properties. For Hume, an object is a set of properties, some of which are instantiated for some period of time. Enduring objects as well as (quasi) instantaneous events are particulars since they are both instantiations of properties in some spatiotemporal region.

As we readily see, this first definition of causality is based on the observed repetition of the succession of certain kinds of events characterised by similar relevant properties, *without involving any necessity*. In contrast, the second definition of causality, which Hume calls "natural", and which can also be called *subjective, mental* or *psychological*, involves the notion of necessary determination.

> A cause is an object anterior and contiguous to another, and so united to it in the imagination that the idea of the one determines the mind to form the idea of the other, and the impression of the one to form a more vivid idea of the other. (Hume 1978, p. 170)

For Hume, an impression is a *sense datum*, a sensation, for example of blue, such as the one we have when we look at a cloudless sky on a clear summer day. An impression is subjective and mental. An idea is, for Hume, a weakened impression, like the blue sensed in a previous experience and remembered on a grey winter day. In this way, we form in our imagination the idea of blue. Returning to Hume's example, since we have observed fire and seen it constantly and repeatedly followed by smoke, the force of *habit* irresistibly leads us to infer the presence of smoke when we see fire, or conversely to infer the presence of fire when we see smoke. In other words, even though we now (supposedly) have only the sensory impression of smoke, without actually seeing the fire, our mind is necessarily determined to form the more vivid mental idea of fire.

Our concept or idea of causality is for Hume a *natural* (necessary) relation, which is rooted in the natural workings of the human mind. Since no sensory impression corresponds to this relation, its necessity lies exclusively in our mind. Such a necessity rests solely on the causal inferences we make under the constraint of habits acquired as a result of endlessly repeated observations of associations of types of events. We certainly cannot refrain from projecting this necessity into reality and believing that an external necessary causal connection exists. But this belief is not grounded in experience and is therefore, for Hume, unjustified. It follows from this that the only causality that experience consents us to call "real" is the one which satisfies Hume's first definition. Real "philosophical" causality is a merely contingent relation between events.

As a consequence of the regularist conception of causality defended by Hume, his disciples, whether nominalists or not, hold that causal laws are contingent. Also, most neoregularists hold that laws are propositions and do not include laws in their ontology. Most but not all. Stathis Psillos for example holds that "a law of nature is a regularity characterised by the unity of a *natural* pattern." (Psillos 2014b, p. 24) For clarity, the philosophers who adopt a propositional view of laws should speak of "nomological propositions" instead of laws, in order to avoid confusion with the defenders of ontological views of laws, such as Psillos. However, in my discussion of the neoregularist conception I will use the word "law" to mean a nomological proposition, for the sake of brevity.

For regularists, scientific laws like Newton's second law, are universal propositions that describe contingent regularities, which are recurrent kinds of processes whose events stand in causal relationships, in the sense of Hume's philosophical definition of causality. In another logically possible world, a massive body might not behave in accordance with Newton's laws. This also applies to non-causal laws, such as the Boyle-Charles law.

At this point, the standard distinction between two kinds of necessity with regard to the truth (and falsity) of propositions must be recalled. The first kind of necessity has to do with the necessary truth of logical laws and true analytic propositions, and also logical and analytical contradictions (which are false in all possible worlds). This first kind is often called *de dicto* (literally, about what is said) because it is relative to our language. For strict empiricists, only analytical propositions[23] are necessary. On the other hand, the truth (or falsity) of factual propositions is always contingent. For regularists, there cannot be *de re* (about the thing) necessity. There are no facts that make some propositions necessarily true.

If a proposition is a law, its truthmakers are regularities which always occur, under the proviso that no disturbing factors interfere. This *proviso* for the truth of a law is expressed by the "ceteris paribus" or "all other things being equal" clauses (I will return to this tricky issue in Chap. 5). A law remains approximately true in circumstances sufficiently similar to those in which the law has previously been verified. Of course, the falsification of a proposition believed to be a law cannot be ruled out. If a universal proposition comes to be falsified, it can obviously no longer be considered a law.

Falsifiability is distinct from contingency. Certainly, if a proposition is necessarily true, it cannot be falsified. But we could be mistaken in believing that a proposition is necessarily true. For the regularist, a law that is true everywhere and always remains a non-necessary truth. However, given our epistemic limitations, we can never be certain that a regularity observed so far will continue to occur everywhere and always. The regularist is a fallibilist. In practice, however, the falsification of a law will usually result in restricting its truth to the empirical domain in which it has extensively been verified before. Today, we have excellent reasons to believe that the fundamental laws of classical mechanics remain true, to a very good approximation, when velocities are much smaller than the velocity of light,

[23] Logical propositions which are necessarily true or false are analytical since their truth or falsity depends on the meanings of logical constants.

gravitational fields are weak and quantum restrictions can be neglected,[24] which is the case in most ordinary contexts. This is why the classical laws continue to be taught today – as it should – in schools and universities and are still used by engineers.

The main problem facing the regularist account of laws is what Bas van Fraassen called the "problem of identification":

> (...) one must identify the relevant sort of fact about the world that gives "law" its sense. (van Fraassen 1989, p. 39)

In other words, we need to identify the kind of facts that make the statement "'p' is a law" true (where 'p' refers to a universal statement). The problem of identification has two aspects. The first concerns the formulation of criteria that enable us to recognise laws, i.e. nomological propositions, and to distinguish them from universal propositions whose truth is purely accidental, that is, the result of chance or fortuitous circumstances. This difficulty, which is *epistemic*, is a classical problem with which all empiricists have struggled since Hume. The challenge here is to answer the following question: how, from perceptual experience alone, can we identify nomological regularities and avoid attributing nomicity to regularities which are merely fortuitous?

The second aspect of the identification problem is *ontological*. What is the sort of "fact about the world", as van Fraassen writes, which endows some propositions with their nomicity status? It is paramount to realise that the truth conditions of the statements 'p' and "'p' is a law" are not identical. The truthmaker of 'p' is a regularity. But the truthmaker of " 'p' is a law" or, equivalently, the statement "It is a law that 'p' " is not a regularity but another kind of fact whose nature must be made explicit. The regularist is required to tell us what makes the statement "'p' is a law" true, or in other words to identify the ontological grounds of the nomicity of 'p'.

Let us look at the *epistemic* problem first. What criterion would allow us to recognise a genuinely nomological proposition, without running the risk of attributing the status of law to a universal proposition, true of course, but whose truth is purely accidental? Humeanss are required to overcome this difficulty by proposing criteria capable of distinguishing a nomological statement such as "All objects made of uranium 235 have a volume of less than one cubic meter"[25] from a statement such as the one proposed by Reichenbach: "All gold cubes are smaller than one cubic mile".[26] As far as we can tell, both statements are true. But the first statement is a *bona fide* law, which belongs to the physical theory of nuclear reactions, while the second describes only a purely accidental feature of our universe. To solve the problem of sorting out nomological propositions from accidental generalisations, empiricists like John Stuart Mill (1843) and Frank Plumpton Ramsey (1928)

---

[24] When the physical quantity of "action" (equal to the product of energy and time) is large, the quantum effects are minimal.

[25] If a quantity of uranium 235 exceeds its 52 kg critical mass a nuclear chain reaction is triggered.

[26] Reichenbach (1947, p. 368)

underlined the *functional role* of laws in the *context of a theoretical system of deductively organised propositions*. In doing so, they paved the way for the elaboration of a highly sophisticated view that culminated in the neoregularist conception of laws developed by David Lewis (1973, 1983) and called the "Mill-Ramsey-Lewis (MRL) conception" or "systematic view".

According to David Lewis (1973, p. 73), "a contingent generalisation is a law if and only if it appears as a theorem (or axiom) in each of the deductive systems that achieve the best combination of simplicity and strength." A theory is all the *stronger* when it allows the deduction of a greater number and variety of empirically correct predictions, that is to say when it contains more accurate information on the phenomena in its domain. Also, we like *simple* theories better, and we are not inclined to award the title of theory to lengthy compilations of observations (such as the astronomical tables or almanacs mentioned by Lewis). Actually, such compilations hardly deserve to be called "theories". For convenience, we construct theoretical systems that contain universal propositions – the axioms – from which one can deduce a potentially infinite class of propositions that describe, for example, the planetary positions at specific times. This is more economical and simpler. But the concern for simplification most often results in a loss of informative content. The law for the simple pendulum that does not mention friction is simpler but less accurate than the law which contains an additional term which takes friction into account.

The *desiderata* of simplicity and strength pull in opposite directions: what we gain in simplicity, we lose in strength, and vice versa. For this reason, Lewis recommends choosing the most *balanced* axiomatic systems, i.e. those that achieve the best compromise between the requirements of strength and simplicity. This is in fact what scientists do. But most of the time, there is no *single* axiomatisation that achieves the best equilibrium between these two requirements. Several deductive systems may comply with this requisite of equilibrium. This is why Lewis defines laws as the propositions *common* to *all* deductive systems that achieve the best balance of simplicity and strength.[27]

It should be emphasised that, according to the best system criterion proposed by Lewis, an isolated proposition can in no way be a law. A proposition must be integrated in a whole system, namely an axiomatized theory complying with demanding conditions, in order to be identified as nomological. Lewis thus claims that his neoregularist approach solves the problem raised by *vacuous* laws, i.e. the ones that are not made true by any regularity, such as, according to him, the Newtonian law of inertia. For Lewis, if the presence of such laws contributes to achieving the best balance between simplicity and strength of an axiomatic edifice, they fully deserve to be called "laws".

[27] In the – admittedly rare – eventuality of the availability of theories that are empirically equivalent with respect to *all* possible experiences, i.e. equivalent in the strong sense proposed by Hans Reichenbach, and which also satisfy the systematic criterion proposed by Lewis, the laws are identified as the universal propositions common to these equivalent theories.

Let us now turn to the problem of identification in its *ontological* aspect. If we want to preserve the core of Humean empiricism – which is no doubt the objective pursued by Lewis – laws are universal propositions which describe *contingent* regularities. The facts of the world that are supposed to make laws true are observed regularities. But if what makes a law 'p' true is a regularity, how can a vacuous law have a truthmaker? No empirical regularity accurately verifies the law of inertia.[28] Moreover, what confers to a statement 'p' the status of law and makes the statement "'p' is a law' "true, is a fact about an axiomatic system (or a class of such systems). This is not a fact about the world, but a fact that can be called "theoretical", since it consists in 'p' being included in a theoretical system that satisfies the conditions of simplicity, strength and balance between them. Simplicity is a subjective or perhaps pragmatic requirement that does not necessarily have a correlate in reality. Lewis does not specify (neither does he believe that he has to) what fact in the world makes one axiomatic system more satisfactory than others according to his criterion of "best balance of simplicity and strength". The difficulty here is not to identify the reality that makes a universal statement 'p' true, but that which makes the statement "'p' is a law" true. Now, whether a proposition belongs to a theory having the internal characteristics stated by Lewis does not seem to be a fact of the real world, but a feature relative to the theories we elaborate.

In addition to its compliance with empiricist requirements, the neoregularist conception can be praised for providing a straightforward solution to the *problem of inference* (van Fraassen 1989, p. 39). According to any satisfactory conception of lawhood, the statement "'p' is a law" must imply 'p', where 'p' is a universal statement whose truthmaker is a regularity. Since for the MRL conception, a law is a universal proposition which belongs to a theory satisfying the conditions stipulated by Lewis, this makes the solution of the inference problem trivial.

4.4 The Difficulties of the Regularist Conception

Numerous scholars (Armstrong 1983; van Fraassen 1989; Carroll 1990; Ellis 2001; Psillos 2002; Dorato 2005; Bird 2007; Schrenk 2017) have criticised the neoregularist conception of laws, shedding light on its significant shortcomings. Foremost among these criticisms is the limitation of Lewis's epistemic criterion of nomicity, the application of which is restricted to theories possessing an *axiomatic formulation*. However, it remains a stark reality that only a limited subset of scientific theories can be effectively axiomatized in the sense delineated by Lewis,[29] even in the realm of physics. It is highly doubtful that all theories pertaining

[28] Yet it seems to me that the so-called "vacuous" laws, such as the law of inertia, can be held to be approximately true when friction and other disturbing forces are very small. Such circumstances can be realized in a laboratory for short distances. They thus have truthmakers and are not vacuous.

[29] For Lewis, as for David Hilbert, an axiomatisation is a set of statements organised in a deductive logical system. For Patrick Suppes (2002), on the other hand, axiomatizing a theory is tantamount to defining an ensemblist predicate.

to various scientific domains could be axiomatized. Moreover, when an axiomatisation is available, alternative axiomatisations are always possible in principle. Given that it is not excluded that the axiomatisation of a theory that achieves the best balance between simplicity and strength is unknown to us, assessing the best equilibrium between simplicity and strength for a theory might always lie beyond our grasp. What is more, even if we knew all the possible axiomatisations, the concepts of simplicity, strength and balance between them are not clearly defined and remain hopelessly vague (van Fraassen 1989, p. 42). Last but not least, our understanding and application of the concepts of simplicity, strength and balance between them depend upon our *subjective* way of looking at them. They are in the eye of the beholder. Thus, we are not in a position to claim that these concepts reflect real and objective facets of the world. While the objective existence of regularities is undeniable, the assessment of the informative strength of a theory hinges on our subjective weighing of its correct empirical predictions, accentuating some while diminishing the significance of others.

Consequently, there is no guarantee that the nomicity of a universal proposition, according to the best system criterion proposed by Lewis, is grounded on some reality. The fulfilment of this criterion depends on our human, subjective and pragmatic interest in organising a theory into an axiomatic system in the way which is most convenient for us. The criterion of nomicity championed by Lewis can be deemed to be purely epistemic – even "chauvinistic" (Carroll 1990, p. 202) – that is to say, relative to the current state of our knowledge and cognitive interests.[30]

At the end of the day, the *ontological* problem of identification remains unsolved in Lewis' approach. No external reality capable of grounding nomicity is identified. Once again, we are confronted with an idealistic manoeuvre. By rendering the notion of nomicity subjective, Lewis echoes Hume's ploy of locating causal necessity in our minds. In a manner reminiscent of Hume, Lewis does not anchor the truth of a statement of the form "'p' is a law" in some reality in the world, but rather in our subjective assessment of specific traits of our axiomatic constructions, with no assurance that these traits (or some of them) correspond to aspects of reality. For Lewis, there are no empirically accessible facts in the world upon which nomicity rests.

By liberating the MRL (Mill-Ramsey-Lewis) conception from its subjective aspects, Stathis Psillos' rendition of the neoregularist conception (2002, p. 154) represents a substantial advancement in addressing the ontological problem of identification. Psillos embraces a *realist* view according to which the organisation of a theory (according to Lewis' best system criterion) reflects the *objective global organisation* of factual regularities, that is, the real structure of the world. Inspired by

[30] Anjan Chakravartty pointed out to me that accusing a follower of Hume of subjectivity constitutes a *Petitio Principii*, as nomicity, much like causality, resides in our minds rather than being inherent to the external world. Nevertheless, it is worth noting that if the notion of necessary causation is in us, it is rooted in a shared, universal aspect of human nature. In contrast, the concepts of simplicity, strength and balance between them are subjective, intimately tied to the varying assessments made by individuals or groups of people.

Ramsey (1928),[31] Psillos calls this organisation the *web-of-laws*. Despite the intricateness of the concepts involved in Lewis' criterion (simplicity, strength, balance) that resist precise characterisation, Psillos' proposal is quite appealing. If "*the world has an objective nomological structure*" (Psillos 2002, p. 154), this fact about the world – as a whole – makes statements like "'p' is a law" true. Consequently, the nomicity of laws is ontologically grounded.

However, is such a solution to the ontological identification problem acceptable within the framework of the empiricist philosophy Psillos tries to articulate? The organisation of the world as a whole can hardly be considered an observationally accessible fact. We obviously only have observational access to parts of the world whose extent in space and time remains very limited. Therefore, as far as the *epistemic* problem of identification is concerned, the web-of-laws posit does not offer significant progress with respect to Lewis' solution.

It is indeed plausible to argue that the organisation of a theoretical system has a counterpart in empirical reality. Yet such contention offers limited assistance when it comes to deciding in practice whether a given theoretical system is better than another. Given that we lack empirical access to the totality of this presumed real counterpart, there exists no strong empirical basis to believe in the existence of such a comprehensive global organisation of the world. Thus, we find ourselves facing the same predicament as Lewis did. The unavailability of empirical criteria for the epistemic identification of laws obstructs the identification of any external reality capable of underpinning their nomicity and, thereby, the resolution of the ontological problem of identification.

Another crucial objection to the neoregularist account requires discussion. Even if laws could be identified by the Mill-Ramsey-Lewis criterion of nomicity, they would fail to imply the truth of counterfactual conditional statements. A counterfactual statement with the logical form "If p, then q" contains a false antecedent p (contrary to facts) and a consequent q which describes what would happen if the antecedent were true. The statement "If this metal bar were heated, then it would expand" is a (presumably true) counterfactual. Counterfactuals are subjunctive conditionals.

A widely accepted requirement for an adequate philosophical theory of lawhood is that laws entail counterfactuals. If the statement "All metal bars expand when heated" is a *law*, it should entail the counterfactual "If this metal bar were heated, then it would expand". However, this counterfactual is not true due to its false antecedent (by hypothesis, the bar is not heated); this would be the case if we were dealing with the material implication of classical extensional logic.[32] In a counterfactual statement, the link between the antecedent and the consequent is stronger and incorporates a form of necessity. If the bar were heated, it would *necessarily* expand.

---

[31] Such claim reminds us of what Wittgenstein said in the *Tractatus*, at least as far as simplicity only is concerned. "We are (. . .) told something about the world by the fact that it can be described more simply with one system of mechanics than with another." (Wittgenstein 1971, 6.342)

[32] According to the truth table for material implication (written $\supset$ or $\rightarrow$), $p \rightarrow q$ is true if the consequent q is true or if the antecedent p is false. The material implication is false if an only if the antecedent is true and the consequent false.

A counterfactual statement asserts a necessary link between the antecedent p and the consequent q. This is why a counterfactual is symbolised by p □ → q where □ is the "box" or "square", which is the symbol for necessity in modal logic.[33]

David Lewis endeavours to account for the truth of counterfactual conditionals by resorting to the similarity between logically possible worlds (Lewis 1973). A possible world is similar to ours if it satisfies the same laws as those of our actual world. The fewer and smaller the differences in initial conditions, the closer are similar worlds one to the other. In this framework, the assertion that Napoleon would have triumphed in the Battle of Waterloo had the Prussians not arrived in time is true if, in the world most similar to ours with respect to the relevant facts, the Prussians arrive late, and Napoleon emerges victorious. According to Lewis, these facts, which actually occur in the possible world closest to ours, constitute the truthmaker of this counterfactual statement. Identifying laws is thus pivotal for establishing the truth conditions of counterfactuals.

To ground the truth of counterfactuals, Lewis postulates *the concrete existence* of possible worlds, a postulate that carries significant metaphysical weight.[34] Such a postulate is supposed to be justified by its explanatory force or, as Lewis says, its "theoretical utility". It seems paradoxical that, in an effort to preserve Humean contingency, Lewis is compelled to postulate the existence of facts in worlds that are – as a matter of principle – empirically beyond our reach. I don't disregard the considerable value of resorting to possible worlds in the realms of logic and semantics. But I don't see why we should include a plethora of possible worlds in our ontology, deeming them as real as our actual world. In Chap. 5, I will try to show that the nomicity of certain universal statements, as well as the truth of counterfactual statements, can be grounded in causal powers. The reality of these powers – within our world – can be established, despite their unobservability, on the basis of ordinary experience. Such an approach avoids the necessity of resorting to possible worlds and the contentious notion of resemblance between possible worlds.[35]

Compounding the challenges of the Mill-Ramsey-Lewis approach, numerous critics remarked that the simplicity of an axiomatic system depends upon the choice of the *predicates* employed for formulating the axioms.[36] To circumvent this difficulty, suggestions have been made to formulate theories by eschewing the

---

[33] On attempts to provide extensional (non-modal) semantics of counterfactuals and their failure, see Schrenk (2017, Chap. 3).

[34] Interestingly, Lewis replies to the charge of blatantly violating Occam's razor thus: "You believe in the actual world already. I ask you to believe in more things of that kind, not in things of some new kind." (1973, p. 87) The distinction between quantitative and qualitative parsimony in this context is carefully analysed by Sendlak (2018).

For the difficulties faced by the regularist conception of laws in relation to the support of counterfactuals, look also at Armstrong's illuminating discussion of this issue (Armstrong 1983, pp. 46–52).

[35] For a critique of his conception of similarity between possible worlds, see Psillos (2002, pp. 94–96).

[36] Psillos (2002, pp. 154–155) again offers a clear presentation of this issue.

utilisation of "unnatural" predicates such as "x is H if and only if x is F or G". As discussed earlier, these predicates, often deemed artificial and unnatural, fail to correspond to real properties in nature.

How can we effectively identify natural predicates? These predicates should denote genuine natural properties, without resorting to the very notion of nomicity, which would ensnare us in circular reasoning. Lewis is well aware of this and has devised several strategies to overcome this hurdle. One such approach, which is more aligned with empiricist orthodoxy, starts by examining the observed similarities among particular entities. Predicates like "cubic" or "spherical" refer, according to Lewis, to natural properties because they are instantiated in a number of particulars as substantiated by observations.[37]

Undoubtedly, observation serves as a good indicator of property instantiation. Yet given our interest in the truth of scientific laws, we must establish the existence of the properties mentioned in these laws. As advocated in the preceding chapter, we are entitled to believe in the reality of properties solely when these are directly or indirectly observed. In contrast, belief in the existence of purely theoretical properties is never justified. In consonance with scientific realism, natural properties are thus identified with the properties successfully referred to by natural predicates in scientific laws.

Notice that realism concerning properties is compatible with various ways of delineating natural classes – those sets of particulars sharing identical natural attributes. The world can be carved up in different ways, without generating conflict with realism. We remain free to organise the world on the basis of the standpoints of physical, chemical, biological, economic, sociological, and other theories as long as we avoid simultaneously assigning incompatible properties to the same particular entity.

As said before, my contention remains that universal propositions belonging to a scientific theory cannot be identified as laws unless convincing reasons support belief in their (approximate) truth. However, from the vantage point of neoregularists, a theory may contain universal statements mentioning predicates intending to refer to purely theoretical properties. Such statements may contribute to render the axiomatic formulation of the theory a best system in Lewis's sense. Even vacuous laws can contribute to attaining the best equilibrium between simplicity and strength. Should they fulfil this role, they fully deserve to be awarded the title of scientific and natural laws. However, they obviously lack truthmakers. If adherents of regularism regard them as true, their concept of truth cannot comply with the correspondence view of truth. Furthermore, empirical evidence in favour of such truth claims remains utterly beyond our reach.

[37] The difficulties encountered by Lewis' treatment of natural properties are examined by Carroll (1990, pp. 199–202).

4.5 The Categorical Necessitarian Conception of Laws

In the face of the challenges raised by neoregularist contingentism, the only viable path forward is the pursuit of a necessitarian account of laws. Laws are necessary, at least in our actual world, in a sense that must be made clear. In order to solve the ontological problem of identification, necessitarians are required to specify the nature of the realities that ground this necessity. Furthermore, these realities will also have to be empirically accessible, otherwise we would lack a convincing rationale to believe in their existence.

David Armstrong (1978), Fred Dretske (1977) and Michael Tooley (1977) have developed a necessitarian conception of laws (referred to by the acronym "ADT") exclusively based on categorical properties. According to them, the predicative terms in scientific laws denote categorical properties or dispositions reducible to categorical properties. On this, they are in full agreement with neoregularists who also adamantly support categorical monism.

Despite their disagreement on some issues, Armstrong, Dretske and Tooley all argue that the statement "'p' is a law" is true if 'p' describes a *categorical relation of necessitation between categorical properties*. According to them, a law is the instantiation of a relation of nomic necessitation between real properties. Thus, categorical necessitarians advocate an ontological rather than propositional conception of laws.

Consequently, according to the ADT conception, a particular which instantiates some specific properties will also necessarily instantiate certain other properties if the latter are related by nomic necessitation to the former. The regular behaviour of some particulars is underpinned by a law which impose or govern such regular behaviour. If a relation of necessitation obtains between, on the one hand, the (complex) property of being a massive body and being subjected to a force and, on the other hand, the property of being accelerated, then – necessarily – all massive bodies subjected to a net force will manifest an acceleration (under proper circumstances), as stated in Newton's second law. This brings us back to the classical view, championed among others by Descartes, that laws *govern* natural processes, in the way that a lawgiver god would govern the world, while removing, however, any moral or theistic connotation from this governance.

On pain of inconsistency, the nomic necessitation relation must be a categorical property. In no case can it be a (irreducible) power or disposition. Moreover, such necessitation relation is a second-order relation since its relata are properties, which are first-order entities. In the usual formulation of laws as universal statements, only categorical properties are referred to. What *makes* a nomological statement *true*, such as "All metals expand when heated",[38] is a universal regularity (given *ceteris*

[38] For the sake of completeness, let me mention that metals dilate according to the nomological formula $\Delta x = k\ \Delta T$. The variation in length is proportional to the variation in temperature, according to a factor k specific to each metal. This law is a consequence of the kinetic theory of solid bodies.

paribus conditions). Yet what makes this statement nomological is the instantiation of a necessitation relation N between the universal properties denoted by "being a heated metal" and "expanding". If we take two categorical properties denoted by "F" and "G", the statement "N (F, G)" is supposed to imply the statement "All things that have the property F also have the property G". The latter statement is then necessary as well. Yet its necessity is conditional because it depends upon the existence of a *singular* fact, namely the instantiation of a necessitation relation between universal properties, which are, according to ADT, categorical universals.

Thus, for Armstrong, Dretske and Tooley, *laws are facts*. N (F, G) is a second-order fact (Armstrong 1983, p. 85). It is, as it were, a *nomic fact*. According to the ADT view on laws, the universals F and G are necessarily related in our world, and perhaps also in some other logically possible worlds, but *not in all* of them. In some other possible worlds the universals F and G do *not* stand in the relation N. In some possible worlds, the universal F is necessarily related to another universal H, but not to the universal G. The fact that the properties F and G are related by N is then a fact relative to some worlds, but not all, and thus is a *contingent* fact. Although necessary in our world, the laws of nature are radically contingent, and on this issue the ADT account concurs with the Humean view. The necessity of laws is not absolute, i.e. holding for all possible worlds; is only relative to a subset of them.

This may seem paradoxical, but such a view allows necessitarian categorists to distinguish nomological statements from *logical* laws. The latter are true in *all* possible worlds, and therefore absolutely necessary, whereas the laws of nature are true relatively to some of them. Nomic, or natural, necessity is *weaker* than logical necessity. A law of nature that is necessary in our world (and some others) may no longer deserve such title in some other possible worlds. For example, in some other worlds, the universals denoted by the predicates "being a free body" and "being in uniform rectilinear motion" are not connected in a necessary way. In some worlds, free bodies would be (for example) at rest or moving in uniform circular motion, contrary to the law of rectilinear inertia holding in our world.

The natural necessity advocated by Armstrong, Dretske and Tooley is a necessity *de re*, not *de dicto*. Assuming that the predicates "having a mass" and "having a velocity" refer to the same properties of mass and velocity in all possible worlds,[39] Newton's first law is false in a world where the trajectories of free bodies are circles (whose radii would depend on the initial conditions) travelled at constant speed. In such a world the properties of mass and velocity have different causal roles from those they play in our world.

[39] Such assumption is far from being evident, since categorical properties are taken to be quiddities according to the proponent of the ADT view (see below).

4.6 The Difficulties of the Categorical Necessitarian Conception

The supposition that properties can have distinct causal roles in different possible worlds raises a first difficulty: how is the identification of the *same* property achieved across possible worlds? If property F is not necessarily related to the same properties in all possible worlds, its causal role varies according to the world in which it is instantiated. In worlds distinct from our actual world, massive bodies might repel each other. Can we still claim that the property of gravitational mass in that world is the same as the property of mass in our actual world? Categorists answer that it is. But in metaphysics, as elsewhere, you don't get anything for nothing. And the price to pay is high. One is forced to accept that the identification of a property across possible worlds rests on a postulate; one simply posits that the *same* property may have different causal roles in other possible worlds. Conversely, different properties in our world might have identical causal roles in other worlds. This position is called "quidditism" (from the Latin "*quid*", "what?"). According to this view, each universal has a primitive identity of its own, namely, a "quiddity" that is independent of its possible causal roles, which may vary in different possible worlds. As Armstrong puts it:

> (...) natural properties have a nature of their own, and it is at least metaphysically possible (...) that the same properties are associated with different causes and effects, that different properties are associated with the very same causes and effects (...) (Armstrong 1999, pp. 26–27).

However, it is hard to get a clear picture of what it is for a property to have a "nature of its own".[40] The identity of a property would be based on its own, primitive nature, of which nothing can be said, which is then shrouded in deep mystery. Quidditism is the counterpart, for properties, of haecceitism for particulars. "Quidditism is to properties what haecceitism is to individuals." (Lewis 2008, p. 209) According to haecceitism (from the Latin "*haec*", "that"), each particular is simply identical to itself; it has a primitive identity.

While rejecting haecceitism for particulars, Armstrong is compelled to adopt quidditism for properties. Within a world, such as ours, natural properties confer powers on their bearers (Armstrong 1978 Vol. II, pp. 43–47). Properties lacking a causal role could not interact with our sense organs and would be unknowable:

> For it is only in so far as properties bestow powers that they can be detected by the sensory apparatus or other mental faculty. (*Ibid.* pp. 44–45)

We have no reason to believe in the existence of entities that lack powers. This is what Armstrong says when he embraces the "Eleatic principle", defended by the Stranger in Plato's *Sophist* (247d-e), and which Armstrong formulates as follows:

---

[40] For an in-depth discussion of the difficulties of quidditism, see for example Mumford (2004, pp. 149–152) and Bird (2007, pp. 73–79).

> (...) everything that we postulate to exist should make *some sort* of contribution to the causal/nomic order of the world. (Armstrong 2004, p. 37)

As a consequence, it seems that there is a contradiction between quidditism and the Eleatic principle. If quidditism is correct, properties have a primitive identity which is not grounded on causal powers. Therefore, primitive identity is ruled out by the Eleatic principle according to which powers are the *mark of being* (Armstrong 1978, Vol. II p. 46). Whether properties have a primitive identity or not doesn't make any difference in what is going on in the world, namely our world, which is the only one that is epistemically accessible to us.

For ADT, powers are ontologically grounded on the laws of nature, and more precisely on the relation of necessitation N, which connects universals F and G. Being categorical, universals are not powers by themselves; they receive their potentiality from outside so to speak, that is, from nomic facts such as N (F, G). Remember: only categorical properties exist; no property is intrinsically powerful. Armstrong refuses to include powers in his ontology. If categorical properties confer powers on their bearers, it can only be by virtue of a relation external, and not internal,[41] to them. In accordance with the Eleatic Principle, the instantiation of a nomic necessitation relation between properties has an impact on what happens in our world.

This brings us to the inference problem that any satisfactory account of laws must be able to solve. The necessitarian categorist has to show how the statement "N (F, G)" entails the universal statement "(x) (Fx →Gx)" which describes a regularity. The latter statement can be rewritten as "(x) (R (Fx, Gx))" where "R" denotes a relation of regular succession between the instantiations of F and the instantiations of G. But the relation N is by itself categorical and passive. In order to guarantee the necessary occurrence of the regularity, we have to resort to a new relation N′ (N, R) which would bestow a causal role on the relation N with respect to R. As N′ is in turn a categorical (third order) relation, we are drawn into an unwanted infinite regress (see Bird 2007, p. 94). It is hardly surprising, after all, that categorical properties are unable to confer modal force to other categorical properties. Is there a way out of this quandary? To answer that question, we must examine how the ADT view endeavours to solve the ontological problem of identification and the problem of inference.

At first sight, the ontological problem of identification is solved. The fact of the world that grounds the nomicity of a law is the existence of a relation of necessitation between properties. But this is not enough. We need to say more about this relation of necessitation. To solve the problem of inference, the necessitation relation must be identified in such a way that the regular successive instantiations of properties (*Fa* then *Ga*, F*b* then G*b* etc.) follow from it.

[41] The relation between being red and being coloured, for example, is a necessary internal relation (all red objects are necessarily coloured). In the same way, the relation between a power and its manifestation is also an internal relation (see Chap. 5).

To do this, let us take a closer look at what a universal is. No doubt, a universal is a property that can be instantiated in an indefinite number of particulars. But we must go further and ask ourselves whether these properties are substances, that is, entities that exist by themselves (i.e., that exist *ante rem*) independently of their instantiations, or whether, on the contrary, they exist only insofar as they are instantiated in particulars (i.e. exist only *in re*).

Dretske and Tooley deem universals to be similar to Plato's Forms or Ideas: universals are substantival entities which exist in a transcendent world, distinct from the world of ordinary entities as well as the entities mentioned in scientific theories. If so, instantiations of necessitation relations between properties are facts which are not empirically accessible. What we observe are the particulars that instantiate properties, but not the properties themselves. When we observe a red particular, we see that it is red, but we do not see redness independently of its instantiation in the particular. Remember what the sophist Antisthenes replied to Plato: "I see a horse, but not horseness."[42] By postulating a relation of necessity between substantial properties, Dretske and Tooley decisively enter the territory of metaphysics.

As Antisthenes reminds us, we have no perceptual access to properties that are not instantiated. This is why Dretske and Tooley rely on the good old explanationist strategy. They justify belief in the existence of substantial universals by invoking their explanatory power with respect to observed similarities. If red objects are similar in colour, it is because they instantiate the *same* substantial property of redness. So far, their argumentation is similar to Armstrong's. But they add that their Platonic conception, unlike Armstrong's, has no difficulty regarding the existence of properties which are not instantiated at any instant of time, which could be the case for determinable properties. The Platonic account straightforwardly provides truthmakers for nomological statements that mention determinable properties, some values of which are not instantiated. As for the universal N – which is a relational property instantiated in universals – we know (according to Dretske and Tooley) that it exists because it allows us to (better?) *explain* the observed constant conjunctions of certain pairs of universal properties. At any rate, explanationist strategies do not consent to infer realities, as argued before. What is more, since we cannot perceive transcendent, uninstantiated properties, Dretske and Tooley are left without a satisfactory solution to the *epistemic* problem of identification.

Suppose now, for the sake of discussion, that such Platonic universals exist and let us examine whether neoplatonists manage to solve the ontological problem of identification and the problem of inference. Let us start with Tooley's definition of the necessity relation N:

Definition: N is a contingent, irreducible, second-order relation of necessitation between universals of order one, such that the constant conjunction of instantiations of these universals is a *consequence* of the assertion that these universals stand in the relation N.[43]

---

[42] Attributed to Antisthenes by Simplicius in his *Commentary on Aristotle's Categories*.

[43] This definition is a simplified version of the one given by Tooley (1977, pp. 49–50).

Nomic regularities depend on the existence of the necessity relation N between categorical universals. On the other hand, accidental regularities do not depend on the existence of such a necessitation relation between universals. This makes it possible to ontologically distinguish between accidental and nomological regularities. The *ontological* problem of identification is resolved!

But this is an illusion. The possibility of deducing the co-instantiation (or successive instantiations) of properties from the fact that they stand in the relation N is part and parcel of the definition of N. Without this the relation N is powerless to ground the regularities which can be counted as nomological. A solution to the ontological problem of identification must be at the same time a solution to the problem of inference. The two problems must be solved together or not at all. At this juncture, the challenge is to show how one can infer a connection between properties instantiated in particulars from a relation between properties.

Actually, it is far from obvious that the supposition that there is a relation N between the properties F and G, described by the statement "N (F, G)", implies the co-instantiation (or successive instantiations) of F and G. We must be careful and avoid confusing the substantial property F with the particulars in which it is instantiated. If F is instantiated in the particular *a*, we have a fact F*a*, which is distinct from the transcendent existence of F. We cannot therefore deduce the statement "N (F*a*, G*a*)" from the statement "N (F, G)". A relation of necessitation between transcendent properties does not entail that the *instances* of these properties also stand in this relation of necessitation.[44] *A fortiori*, by generalizing, from "N (F, G)" we cannot deduce "(x) N (Fx, Gx)" which would directly imply "(x) (Fx →Gx))".

Dretske and Tooley concede that the validity of such an inference cannot be proven. This is why Dretske offers the following analogy:

> The legal code lays down a set of relationships between the various *offices* of government, and this set of relationships (. . .) imposes legal constraints on the individuals who occupy these offices (...) Natural laws may be thought of as a set of relationships that exist between the various "offices" that objects sometimes occupy. (. . .) Once an object occupies such an office, its activities are constrained by the set of relations connecting that office to other offices and agencies: it *must* do some things, and it *cannot* do other things. In both the legal and the natural context the modality at level *n* is generated by the set of relationships existing between the entities at level *n* + 1. Without this web of higher-order relationships there is nothing to support the attribution of constraints to the entities at a lower level. (Dretske 1977, pp. 31–32)

This reasoning does not go through, however, as van Fraassen showed (1989, p. 99). If an individual who occupies an office feels obliged to act in accordance with the law, it is because he or she has *previously decided* to comply with the law, and not by virtue of logical coercion. Civil legislation as such does not prevent transgression of it. If it did, there would be no need to enforce laws and imposing penalties for those

[44] A similar objection was raised by Aristotle against Plato. We cannot deduce propositions which describe changes, among which regular changes, from relations between universals. See Ross (1971, p. 158)

who do not abide by them... In an analogous way, from a necessitation relation between universals F and G it does not logically follow that a particular *a* that instantiates F will also instantiate G. Formally, it is clear that the statement "N (F*a*, G*a*)" is not deducible from "N (F, G)". Granted, Dretske only proposes an *analogy* between civil legality and natural nomicity. But this analogy is devoid of any demonstrative force. The problem of inference remains unresolved unless one postulates, as seen above, an indefinite sequence of higher-order relations, i.e. a regression *ad infinitum* whose explanatory force is debatable, to say the least...

Unlike Dretske and Tooley, David Armstrong opts for an Aristotelian conception of universals according to which a universal must be instantiated at some place and time, be it past or future. For Armstrong, universals exist *in re*, that is, in particulars, and not *ante rem*, independently of them.

> The Principle of Instantiation for properties asserts that, for each property P there exists (not necessarily now) a particular *x* such that P*x*. (Armstrong 1978, Vol. II p. 76)

Does this principle of instantiation allow him to eliminate the regression above, to solve the ontological problem of identification and by the same stroke the problem of inference? As we will see, this is not the case.

In a short paper, Armstrong (1993) attempts to answer van Fraassen's objections and to show that his version of necessitarian categorical monism does provide a solution to the ontological problem identification as well as the inference problem. The challenge for Armstrong is to empirically identify the relation of necessitation N in such a way that the observed regularities occur by virtue of this necessitation relation between properties. Armstrong had originally argued that the necessity relation was a *primitive* relation. But, like the definitional approach of Dretske and Tooley presented above, this solution was accused of being purely verbal by David Lewis (1983, p. 366), in a rather ironical way. This objection is known in the literature as the "mighty biceps" objection. It is not enough to call the relation N "relation of necessitation" for it to provide *ipso facto* a solution to the problem of inference, just as it is not enough to be called "Armstrong" to have strong biceps... (van Fraassen 1989, p. 98).

This is why Armstrong, at the beginning of his 1993 paper, claims that some necessary singular causal relations are accessible to sensory perception, such as the "pressure on our body."[45] (Armstrong 1993, p. 421) Of course, this is not acceptable to Humean empiricists like van Fraassen, nor to myself, since I do not think that a modal property, such as the necessary relation of causality, is observable, even if this necessity only holds in our actual world. For the sake of discussion, let us admit, however, that some singular causal relations are observed.

Causal relations between singular facts manifest regularities. A fact F*a* is followed by a fact G*a*; F*b* is followed by G*b* etc. We thus have a sequence of facts where causes of the *same type* produce effects of the *same type* as well. As

[45] As early as 1978 Armstrong argued (Armstrong 1978 vol. II, pp. 164–165) that causal relationships could in some cases be directly observed, not inferred, such as perceiving pressure on one's own body when seated.

Armstrong says, “we can infer to the existence of universals to explain these ‘samenesses’.” (*Ibid.,* p. 421) The instantiations F*a*, G*a* etc. of this type of causal *pattern* exemplify the “*very same pattern of universals*” (*Ibid.*). Can we go further and *explain why* this regularity occurs? To work out such an explanation, Armstrong postulates a relation between universals F and G. This relation is the necessary relation of causality C. The fact C (F, G) guarantees, according to Armstrong, that the observed regularity of singular causations occur.

Yet for this to be the case, and for the problem of inference to be solved, we must identify the relation C, which connects *types*, with the observed singular causal relations between *instances* of these types: F*a* causes G*a* etc. (*token-token* causation) Although Armstrong does not explicitly refer to the necessitation relation N in his 1993 paper, it is clear that in his view the causality relation C is identical to N. N is the causality relation between the *types* F, G... which is also claimed to be identical to the relation N between the *tokens* F*a*, G*a*; F*b*, G*b*... And this solves the ontological problem of identification as well as the problem of inference. The statement “N (F, G)” implies “N (F*a*, G*a*)”, “N (F*b*, G*b*)”... and thus (x) N (Fx, Gx) from which “(x) (Fx →Gx)” immediately follows. As Armstrong says:

> (...) if a certain type of state of affairs has certain causal effects, how can it not be that the tokens of this type cause tokens of that type of effect? The inference is analytical or conceptual. (*Ibid.* p. 422)

However, in his 1993 paper, Armstrong elegantly acknowledges that the crucial assumption of the argument, namely the identification of the causality relation between types with the causal relation between their instantiations, is controversial and cannot be proved.

If the relation N cannot be identified with the causal relation between facts (even if we admit that it is observable), the problem of ontologically identifying the nomic necessitation relation N persists, as does the problem of inference. This is the major difficulty faced by the ADT conception of laws.

The same difficulty makes it impossible to satisfy the requirement that a law grounds the truth of a counterfactual conditional proposition. For example, the proposition <if I dropped this pen, it would fall to the ground> follows from the law of gravitation and Newton’s second law. If the inference problem is unsolved, laws cannot logically entail counterfactual propositions: entailments such as N (F, G) → (F*a* □→ G*a*) remain unjustified.[46]

Functional laws raise a further challenge that any conception of laws must address. Most mathematical laws contain functional relations between determinable properties such as pressure, volume, mass, charge etc. The mass of a body, for example, can take an infinite number of determinate values. A very precise value of mass is a property that can in principle be instantiated in an indefinite number of

---

[46]Probabilistic laws, such as the laws of radioactive decay, raise a further difficulty. Armstrong (1983, Chap. 9) discusses this problem in depth, but van Fraassen (1989, pp. 109–116) has shown that the solution he proposes is inadequate. Without going into this technical discussion, I just mention here that I subscribe to van Fraassen’s critique.

particulars; it is therefore a universal. A functional law appears *at first glance* to be an indefinite conjunction of relations of necessitation between determinate values of certain quantities, and thus, an indefinite conjunction of particular laws. Yet it could be that a precise value of mass is never instantiated, with the consequence that a very specific law that relates precise values of mass and force to a precise value of acceleration is not instantiated either. In this case, Newton's second law, for example, would have missing values, empty holes as it were.

For Dretske and Tooley, this problem does not arise since properties do not have to be instantiated to be real. Therefore, laws, which are relations of necessitation between properties, exist independently of the instantiation of these properties and the presence of regularities in nature. However, this Platonic view raises epistemic difficulties that Armstrong's Aristotelian instantiation principle permits to avoid. Then, to solve the problem raised by functional laws, Armstrong proposes that "a statement of an uninstantiated law should be construed as a counterfactual." (1983, p. 112) Statements of uninstantiated laws are statements of nomological singular facts N (F, G) that *would exist* if certain universals F, G, N were instantiated. It is significant that Armstrong speaks here of nomological *statements* rather than nomological facts. Indeed, non-instantiated laws do not exist, in accordance with the instantiation principle. One can then legitimately ask what would be the truthmaker of these counterfactual statements. What supports the truth of such statements, according to Armstrong, is the existence of a law that governs the family of particular laws. Thus, the law $F = ma$ governs the particular laws $F_i = m_i a_i$ (where i is a real number). If precise values F_1 *and* m_1 were instantiated, one would necessarily have an acceleration a_1. Rather than construing a functional law as an indefinite conjunction of particular laws, Armstrong is led to consider Newton's second law as a *higher order* law, that is, as a third-order law. For him, a law is a fact, and I have already said that I prefer not to assign an order to facts. However, orders can be attributed to universals. Since properties such as volume, mass etc. are *determinable* properties, i.e. properties that can have determinate values, Armstrong considers these determinable properties as properties of properties. Determinables are thus second-order properties and the relation of necessitation between them becomes, for functional laws, a third-order relation.

Unfortunately (for Armstrong) we are brought back to the problem of inference: in what way does this third-order necessitation relation N* (F, G) guarantee the necessity of the second-order necessitation relations N (F_i, G_i) for laws that involve determinate values of determinable properties? One can of course simply postulate, as Armstrong does, that N* = N, but the justification of this postulate is problematic since N*and N do not have the same order. What is more, the introduction of properties of properties raises problems of which Armstrong is perfectly aware. It would be fastidious to discuss these problems in detail here (see Armstrong 1983, pp. 114–116). For my part, in addition to being very reticent with regard to the existence of properties of properties (except formal ones), I think that a metaphysics of causal powers makes it possible to address the challenge raised by functional laws, even if we adopt the instantiation principle, as I will try to show in the next chapter.

4.7 Conclusion

In this chapter, after having presented the distinction between categorical and dispositional properties, I have reviewed the conceptions of laws defended by philosophers for whom only categorical properties have their place in an ontology of nature. For categorical monists, all dispositional properties (among which causal powers) are in principle reducible to categorical properties. Categorical monists fall into two broad classes, the regularists and the contingent necessitarians. The regularists are directly inspired by David Hume, for whom the causal relation, in the philosophical or objective sense, is reduced to purely contingent regularities. Under these conditions, it is natural to identify laws with propositions that describe such contingent regularities. To attempt solving the difficulties of a rudimentary regularist conception which is unable to provide a distinction between nomological regularities and purely fortuitous ones, Mill and Ramsey proposed a *neoregularist* view, the most complete version of which has been developed by David Lewis. According to the MRL view, laws are universal propositions that belong to theories organised in axiomatic systems that achieve optimal equilibrium between simplicity and empirical strength. The ontological counterpart of laws, understood as propositions, are the regularities described by these laws. If the neoregularist conception immediately solves the problem of inference, it is less successful when it comes to solving the ontological problem of identification, namely that of specifying "the sort of fact about the world", as van Fraassen says, that grounds the nomicity of laws.

In order to overcome the difficulties of the neoregularist theory, Armstrong, Dretske and Tooley, have developed a necessitarian account of laws that remains faithful to categorical monism and Humean contingentism. According to the ADT view, laws are identified, in an ontological sense, with relations of necessitation between universal properties, i.e. with singular facts. But since this necessity is relative to a given possible world (or a subset of possible worlds), the truth of a nomological proposition remains contingent. As we saw, neither in its Platonic version (Dretske, Tooley) nor in its Aristotelian version (Armstrong), this necessitarian conception succeeds in providing a satisfactory solution to the ontological problem of identification in such a way that the problem of inference would also be solved. Concluding, none of the purely categorical conceptions of laws that have been proposed, whether neoregularist or contingent necessitarian, is acceptable.

Chapter 5
A Dualist Metaphysics of Nature

Abstract I start this chapter with a presentation and critique of Brian Ellis' new essentialism, which consists in a dualist metaphysics of laws based on both categorical properties and powers. Like Ellis, I defend a dualist metaphysics of nature. However, unlike him, I attempt to dispense with essences and natural kinds and advocate an enlarged view of categorical properties. These include not only spatial and structural properties but all observable properties. Besides, I submit that categorical properties are not quiddities and that their identity across possible worlds doesnot reduce to a mysterious primitiveness. I also develop an empirical argumentation in favour of the reality of irreducible causal powers which ground the necessity of laws and their associated counterfactual conditionals. Since the instantiation of powers is relative to some possible worlds, the necessity of laws of nature is contingent. Such conditioned *de re* necessity of laws is distinct from the *de dicto* necessity of analytic propositions.

Scientific realism solves the logical problem of inference and the epistemological identification problem for *scientific* laws. But how are we to identify the ground for the necessity of laws? Granted, we are irresistibly inclined to attribute some sort of necessity to laws. Does this necessity reside only in our minds, as David Hume claimed? Or is it based on external, independent realities?

In the previous chapter we saw that categorical accounts of laws are confronted with serious difficulties. In this chapter, I will argue that resorting to natural powers not only provides an external metaphysical foundation for the necessity of laws but also that belief in natural powers can be empirically justified.

The friends of powers fall into two groups. The first comprises the dispositional monists who, like Stephen Mumford and Alexander Bird, reject categorical properties and contend that all natural properties are powers. The second group includes dualists who, like Brian Ellis and me, advocate an ontology of both categorical properties and powers. Since I concord with Ellis in favouring a dualist metaphysics, I will start with presenting his conception of laws. However, I will not follow him on

M. Ghins, *Scientific Realism and Laws of Nature: A Metaphysics of Causal Powers*, Synthese Library 483, https://doi.org/10.1007/978-3-031-54227-5_5

some important issues. In particular, I will reject his claims that natural kinds are a special sort of universals and that laws of nature are descriptions of the essential properties of natural kinds. I will then proceed to articulate my dualist necessitarian view of laws and respond to some objections.

5.1 The Dualist Essentialism of Brian Ellis

Ellis traces the origin of his essentialism to Aristotle[1] and his metaphysics of natural kinds. Observation attests the existence of entities that show strong similarities to each other, such as humans and more generally members of living species. For Aristotle, animal and plant species are paradigmatic examples of natural kinds. But the particulars belonging to the same species do not look exactly alike. They are not perfectly identical. Although they all instantiate a shared set of properties – which do not evolve in time, according to Aristotle – members of the same species differ slightly from each other in some intrinsic properties. No lion is identical to another. Each lion has some individual properties that singularizes it from all the others. Also, a lion can acquire or lose some of its properties without ceasing to be a lion. A lion without a tail is still a lion. These observations led Aristotle to introduce the distinction between *essential* and *accidental* properties. The essential properties of a lion are the intrinsic properties that are *necessary* – and also *sufficient* – for its membership to the natural kind lion, whereas its accidental properties are contingent, not necessary for its membership to that kind.

Like most authors who accept the existence of natural kinds, Ellis claims that chemical elements and elementary particles provide paradigmatic examples of natural kinds. Each hydrogen atom, for example, has the intrinsic[2] property of having one single proton. This property is necessary and sufficient for something to belong to the hydrogen kind. The kind essence of a thing is the set of properties necessary and sufficient for this thing to be a member of that kind.

> The kind essence of a thing (. . .) is the set of its properties in virtue of which it is a thing of the *kind* it is. (Ellis 2002, p. 12)
>
> (. . .) every natural kind of thing, at every level of generality, has its own distinctive real essence, i.e. its own unique set of intrinsic properties or structures in virtue of which things are of the kinds they are. This is the hypothesis of essentialism. (Ellis 2008, pp. 142–143)

Ellis distinguishes three types of natural kinds. In addition to the traditional kind of objects or substances, which Ellis calls the *substantive kind*, he introduces the natural

[1]Like Ellis, I will not pretend to be entirely faithful to Aristotle's terminology, or to do justice to all the subtleties of his metaphysics (of which I am not an expert).

[2]Ellis defines an intrinsic property as follows: "G is an *intrinsic* property of an object x $=_{\text{déf}}$ x would manifest G in the absence of any accidental force that might affect the properties that would be manifested by G." (Ellis 2001, p. 29) Intrinsic or monadic properties are the properties of the object when isolated. Of course, an intrinsic property could be accidental.

kind of *processes* and the natural kind of *properties and structures*. For the sake of simplicity, I will mainly limit my discussion to issues related to the natural kind of objects or substances.

What is a natural kind for Ellis? First of all, a natural kind K is *not* the set of its members. Ellis is very clear on this:

> For every natural kind, there is an intentionally definable natural kind class, which is the class of its members. The set of all members of this class is the set of instances of the kind. (Ellis 2001, p. 23)

> The class C(K) of instances of K defines what I called the extension of K. (*Ibid.*)

For Ellis, a natural kind is instantiated *by* the members of its extension. A natural kind is a sort of universal. However, it is not instantiated *in* its instances but *by* them. A natural kind is a "quasi-universal" or a "generalised universal". (Ellis 2001, p. 97) A natural kind is what I will call a "*sui generis*" property distinct from the classical or ordinary universals, which are instantiated *in* things, such as having a mass or being pink. As far as substantive kinds are concerned, a natural kind is called by Ellis a "substantive universal". (*Ibid*).

> (. . .) there is a case for regarding the property of being an electron as a conjunctive property, the conjuncts of which are the [essential] properties of electrons. (. . .) However, if there is such a property as this, then it is a very special kind of property. It is instantiated *by* objects (namely, electrons), not *in* them. I call them "substantive universals". (Ellis 2001, p. 97).

To identify the property of being an electron (or electronhood) to the conjunct of its essential properties is plausible for the fundamental constituents of matter since they don't have structural properties. On the other hand, for entities such as the hydrogen atom and horses, which exhibit structural properties, such an identification is problematic for Ellis (2001, p. 73). Yet such identification is not excluded.

> If there is a property of being a horse, then its instances are the things that are supposed to have this property, namely, the horses themselves. So, if there is such a property, then it is, to say the very least, a very peculiar property. (Ellis 2002, p. 46).

Properties such as horseness, electronhood and being a hydrogen atom would be peculiar properties because they are natural kinds (substantive universals in this case) that are instantiated by – and not in – the substances of their associated class.

Let us summarize the main points made above through the example of the electron kind. Electrons are substances with the following essential properties: having a 9.1×10^{-31} kg mass, a 1.6×10^{-19} Coulomb negative charge, a spin ½ etc. The set of essential properties specific to electrons and only them constitutes the essence of electrons. These essential properties are instantiated *in* individual electrons the set of which is the extension of the electron kind. Together, these essential properties determine and make what an electron is. If something does not instantiate one or *a fortiori* several of these essential properties, it is not an electron. An essential property is always an intrinsic property.

The property of being an electron (electronhood) is not an essence. Such property is identified with the natural kind itself, namely the quasi-universal or substantive universal which is instantiated *by* individual electrons. In this sense, a natural kind is *similar* to a universal (it is a quasi-universal) since it is instantiated *by* all its members, although it is not instantiated *in* them.[3] As a way of speaking, the essence of a natural kind is the essence of its members, namely the specific set of their essential properties.

According to Ellis' "new essentialism", as he calls it, the world is hierarchically structured into natural kinds which cut nature at its joints by dividing it according to its natural articulations, in conformity with Plato's metaphor (*Phaedrus* 265e), and this in a *unique* way. The division of nature into natural kinds is in no way arbitrary or conventional. Nor does it respond to pragmatic concerns. This division corresponds to the real hierarchy of natural entities. The natural kinds are clearly and distinctly separated from each other in a *discontinuous* manner. They are, as Ellis puts it, "categorically" distinct.

Moreover, natural kinds can be divided into subspecies by means of natural properties. For example, chlorine (Cl) atoms are members of a natural kind.[4] Their essential property is that all chlorine atoms have a nucleus which contains 17 protons (their *atomic* number is 17). But the nuclei of chlorine atoms can contain 17, 18 or 19 neutrons, for a total of 34, 35 or 36 nucleons (protons and neutrons) respectively. Chlorine atoms can have different numbers of nucleons, i.e. various *nuclear* numbers, and still be chlorine. Having a certain number of nucleons is an accidental property of a chlorine atom.

The natural kind or *species* chlorine has several *subspecies*. The nuclei of the members of each subspecies contain a specific number of nucleons. The chlorine isotopes $^{34}Cl_{17}$, $^{35}Cl_{17}$, $^{36}Cl_{17}$ are three (natural) subspecies of the chlorine natural kind. "(. . .) isotopes are species of the atoms of which they are isotopes." (Ellis 2002, p. 26) Atoms in turn are the constituents of molecules, which correspond to different natural kinds. In this way we have, according to Ellis, a hierarchy of chemical kinds. "The chemical elements [atoms] and chemical compounds all belong in hierarchies." (Ellis 2002, *Ibid.*)

Of course, all members of the same natural kind necessarily have the same essential properties. But members of the same kind may differ from each other in some of their intrinsic properties, as is the case with the isotopes of chlorine. It is not essential for a chlorine's nucleus to contain 18 neutrons: this is an accidental property of chlorine. Yet this property is essential for a chlorine atom to belong to the isotopic kind $^{35}Cl_{17}$. Therefore, it is important to notice that a universal property is essential or accidental *relatively* to the natural kind which is considered. A property is not in itself essential or accidental. It is essential for an electron to be charged, otherwise it is not an electron. An electron cannot lose its charge without ceasing to be an electron. But it is accidental to a crystal whether it is charged or not.

---

[3] We can compare this distinction between the "in" and "by" instantiations with that between the two types of predication distinguished by Aristotle: <Socrates is rational> and < Socrates is *a* human>. The property of being rational is instantiated *in* Socrates while the kind human is instantiated *by* Socrates and not *in* Socrates, according to Ellis.

[4] More accurately, members of a natural kind are members of the class associated with the natural kind chlorine, but Ellis often says that the entities which instantiate a natural kind are members of this kind.

A crystal can be charged or uncharged without ceasing to be a crystal.[5] Likewise, a gas necessarily has a pressure of a determinate value, but it must not have this or that value of pressure in order to be what it is, i.e. a gas.

Necessarily, any member of a natural kind instantiates a distinctive set of essential properties. For Ellis, no possible world contains a chlorine atom which does not have 17 protons. Therefore, the statement that all chlorine atoms contain 17 protons is true in all possible worlds.[6] Of course, some possible worlds do not contain chlorine. But in these worlds, the proposition <all chlorine atoms contain 17 protons> remains true, but vacuously so, according to Ellis. Propositions that attribute essential properties to members of a natural kind are true in all possible worlds. Such propositions are laws of nature and enjoy metaphysical necessity, distinct from logical necessity since their truth is grounded on the essential properties of a natural kind. Contrary to what the proponents of the ADT view claim, the necessity of laws of nature is, according to Ellis, absolute and not relative to some possible worlds.

Some natural kinds have no subspecies. This happens when all members of the natural kind have exactly the same intrinsic properties. As far as we can tell today, there are no two kinds of electrons. There is no intrinsic property that is accidentally possessed by some electrons and not by others. Electrons differ from each other only in their location and their relations to other elementary particles (in their interactions, for example), but these are accidental extrinsic properties. Electrons are members of what Ellis calls an "infimic natural kind". (Ellis 2002, p. 57) By definition, an infimic kind has no subspecies. Contrary to what the term "infimic" might suggest, members of an infimic natural kind need not be simple entities, i.e. entities that are not composed of other entities. Protons and neutrons belong to infimic kinds even though they have parts, namely quarks. The same is true for the isotopes of chlorine, which also are members of infimic kinds, and which contain protons, neutrons and electrons. Thus, we must distinguish between the organisation of nature into natural species and subspecies on the one hand, and the constitution of nature by its most elementary components on the other. Salt (NaCl) is composed of one atom of chlorine and one atom of sodium, but chlorine is not a subspecies of salt. Similarly, protons and neutrons are components of all atoms, but are not subspecies of atoms.

All natural substances are endowed with essential properties that determine what their natural kind is. In this way, nature is organised into a hierarchy of natural kinds with the "global natural kind" at the top, which includes all objects and substances that can exist in the world. (Ellis 2002, p. 36) The infimic natural kinds are at the bottom of the hierarchy since they have no subspecies.

Today, contrary to what Aristotle believed, we know that biological species are not fixed and immutable. Observation of living populations and palaeontological material affords strong evidence in favour of the *fact*[7] of evolution. Over the course

[5]For Ellis, the properties which determine a subspecies (and are therefore essential to them) are not accidental but "incidental" properties (Ellis 2001, p. 78).

[6]Unlike Lewis, Ellis does not postulate the concrete existence of possible worlds.

[7]It is important to distinguish the *fact* of evolution from an *explanation* of this fact, such as a neo-Darwinian explanation.

of generations, biological species change. Some new species come to existence, and this in a *continuous* way, without being possible to determine the moment at which a new species appears and an old one disappears (except in the event of extinction). Since populations of living beings evolve from one another in a continuous way, it is futile to try to identify a set of essential properties the possession of which would be a necessary and sufficient condition for an individual to belong to a given species.

To make his essentialism defensible, Ellis is compelled to embrace a radical position. For him, only the chemical composition of the DNA (deoxyribonucleic acid) of an individual makes it possible to determine the natural kind to which it belongs. Members of a biological kind must have exactly the same DNA. Natural biological kinds do not have subspecies; they are infimic kinds. The members of some species of organisms such as viruses, bacteria, hydras etc. have identical or nearly identical genetic material. But for more complex organisms – plants, animals, humans – their natural kinds, in Ellis' radical sense, usually have one member only. Yet the members of plant and animal species, in the usual sense, do show some resemblance to each other. To account for this indisputable fact Ellis introduces another element in his ontology. A biological species is what he calls a "*cluster kind*" (Ellis 2001, p. 32). The members of a cluster kind exhibit genetical, morphological etc. similarities. Paradoxically, for Ellis, a human individual, for example, belongs to a subspecies of the human – cluster – kind of which it is the only member.[8]

Ellis acknowledges that his metaphysics draws heavily on science, especially particle physics and the classification of chemical elements in the Mendeleev table. His realism is resolutely materialist, and more precisely physicalist:

> (...) every world-object is necessarily a physical object: an object that has energy, and is capable of interacting causally with other physical objects.[9] (Ellis 2002, p. 86)

It is up to the sciences, especially physics and chemistry, to discover the essential properties of natural kinds. Essences can only be known *a posteriori*. Some scientific laws describe the behaviour of the members of substantive natural kinds, i.e. the processes which these members can manifest. For example, when chlorine comes into contact with water, it forms hypochlorous acid (HClO), which is used to disinfect swimming pools. The reaction $Cl_2 + H_2O \rightarrow HClO + HCl$ necessarily – and therefore always – occurs under the right circumstances, as a consequence of the essential *causal powers* of chlorine and water. A causal power of chlorine is the underlying *dynamic power* that necessarily gives rise to this chemical reaction.

In addition to process laws, science contains laws that Chakravartty calls "definitional laws" (Chakravartty 2007, p. 153), such as <Chlorine atoms have 17 protons> and < Water is H_2O >.[10] Necessarily, everything that is water is composed of

[8]Even the DNAs of so-called "identical" twins are slightly different.

[9]In my opinion, such a position cannot be empirically justified, nor can it be proven. It is akin to an act of faith. The existence of entities that are not physical and not cognitively accessible by scientific methods cannot be excluded *a priori*.

[10]Ellis concords with Putnam in asserting that a chemical element of composition XYZ distinct of the chemical composition of our water but having all its surface properties (Putnam 1973), hypothetically present on a twin earth, simply is not water (Ellis 2002, p. 16). Ellis gives ontological primacy to scientific properties.

molecules that contain two hydrogen atoms and one oxygen atom; otherwise, it is not water. Such necessity is not *a priori* but *a posteriori* since the chemical composition of water cannot be known a priori and has been discovered. According to Ellis, it is a law of nature that water has the chemical composition H_2O. If a possible world contains water, its chemical composition is necessarily H_2O. However, there may be worlds that do not contain water. Thus, the existence of water is contingent, although the proposition which describes its chemical composition is necessarily true, as a matter of metaphysical (not logical) necessity: there is no possible world in which this proposition is false. In worlds which do not contain water the proposition <water is H_2O > is "vacuously" true (Ellis 2002, p. 111), but necessarily so.

Ellis champions a dispositional essentialism according to which the members of natural kinds have essential causal powers. Due to its chemical composition, water has the natural power to participate in a number of chemical reactions, described by laws. Similarly, due to its negative charge, the electron has the natural causal power to attract positive charges in conformity with the laws of physics. The necessity of these laws is based on a real causal power – the charge – which is an essential property of electrons. If an entity of a specific substantive kind instantiates a causal power P, it necessarily follows that it will engage in a process described by a law, provided that the right circumstances obtain. This is an important and recurring claim in Ellis's writings:

> (...) a causal power P must have a law of action A that describes what it does when it acts. (Ellis 2012, p. 14)

> For essentialists, laws of nature are explications of the essential properties of the natural kinds. (Ellis 2002, p. 85).

> (. . .) the laws of nature describe the essences of the natural kinds. (. . .) Any analysis of laws of nature must therefore begin with an analysis of the natural kinds. (Ellis 2002, p. 82)

> (...) laws of nature refer to the essential properties of the natural kinds, and (. . .) these are their truth-makers. (Ellis 2002, p. 102)

Ellis thus holds the laws of nature to be explications of properties essential to natural kinds. Laws describe how essential powers are manifested. Laws are *propositions*, not realities. Ellis clearly subscribes to a propositional (rather than ontological) conception of laws. The necessary truth of nomological propositions is grounded on the essential properties of natural kinds. Essential properties such as mass, charge etc. are dispositional, not categorical, properties. The reference to natural kinds such as the electron kind explicitly occurs in the formulation of a law. <All electrons are attracted by positive charges> is a nomological proposition. The manifestation of an essential power in natural kind instantiations is described by a "law of action" (which corresponds to what I have called a "nomological formula"). Such laws are necessarily true, in a strong metaphysical sense, i.e. true in all possible worlds. Again their necessity is not analytical because it is grounded on the reality of powers. The essential causal powers of natural kinds are discovered through scientific activity, and therefore *a posteriori*.

For Ellis, "a causal power is a disposition to engage in a certain kind of process, namely a causal process. A causal process is one of a kind that relates to kinds of events." (2002, p. 48) The first kind of events include all the causes and the second all the effects. Ellis adds that a "specific functional relationship" holds between the causes and effects. Two important points deserve to be highlighted now. First, a power is identified by a kind of manifestation which is described as a functional relation between cause and effect, that is, by a law of action, which is a view I am sympathetic with. Second, Ellis does not discuss the *ceteris paribus* conditions and triggering circumstances in which the causal power "acts". Hence, it is not clear whether the process described by a "law of action" will necessarily happen if the right circumstances obtain. What is guaranteed is only the possibility of such a nomological process and also its regularity for members of the same natural kind. Necessarily, an electron has the possibility to attract positive charges and such necessity flows from the essence of the electron. But Ellis' dispositional essentialism does not provide a grounding for the necessity (if there is one) of the occurrence of a nomological regularity conditioned upon some appropriate circumstances. In other words, Ellis's concept of law does not ground the necessary link between the antecedent and the consequent belonging to some counterfactual conditional. A consequence of this view is that the truth conditions for counterfactuals such as <if a positive charge were present in the vicinity of an electron, it would be attracted by it> are not determined. However, Ellis dodges this problem by claiming that what really matters are the conditions of assertability of such conditionals, not their truth conditions. (Ellis 2002, pp. 131–134) Such position obviously goes against the truthmaker principle.

According to Ellis' dispositional essentialism, causal powers are responsible for the changes happening in the world. These causal powers are real properties instantiated in the members of natural kinds. Being propositions, laws are obviously incapable of *governing* natural processes. The governing business is performed by the causal powers, which make possible specific behaviours for the things in which they are instantiated, whether in a deterministic or probabilistic way. Ellis does not defend an ontology in which laws are irreducible primitive realities. It is thus not quite correct to attribute to Ellis an ontological conception of laws, according to which "laws are just the havings of essential properties by natural kinds" as Mumford says (2004, p. 108). For Ellis, laws are not realities, but propositions that *make explicit* or *describe* the essences of natural kinds, typically their essential powers, whose manifestations are described by laws of action. Like Mumford, Ellis sustains that causal powers are the driving forces that set nature in motion. Natural entities are not passive; they are dynamic and active.

Contrary to what dispositional monists such as Bird (2007) uphold, Ellis' ontology also contains *categorical* properties. These are spatial and temporal properties, such as location, geometric shape, and also structural properties, like the organisation of molecules in a crystal. Agreeing here with categorical monists, Ellis claims that categorical properties are *quiddities*. Without categorical properties, causal powers could not exist. As a matter of fact, the instantiation of a causal power presupposes a location and locations are categorical properties.

> Locations (. . .) must be ontologically more fundamental than the causal powers. For, without them, the causal powers could have no instances, and so could not exist. (Ellis 2012, p. 18)

Moreover, laws very frequently mention categorical properties, such as spatial distance. Although categorical properties are obviously not causal powers, categorical properties are, according to Ellis, "causal factors" (Ellis 2012, p. 22) that contribute to the manifestation of causal processes. The precise way in which a causal power acts depends on the numerical values of categorical properties. For example, the value of the gravitational force between two masses depends on the distance between them. Of course, categorical properties are not modal. But other properties like mass, charge etc. are causal powers which are not reducible to categorical properties. Charge confers on the entity that instantiates it not only the capacity to satisfy the laws of electromagnetism but also the obligation to comply with them if the appropriate circumstances obtain. The necessity of nomological statements depends on the presence in their formulation of predicates that refer to causal powers essential to natural kinds. According to Ellis, the fact that nomological statements also contain predicates that refer to categorical properties does not undermine their necessary truth. (Ellis 2012) (Such claim however raises difficulties that will be examined below.)

Finally, we must say a few words on Ellis' treatment of quantitative properties, the so-called "determinable properties", which he calls "generic properties", such as length, mass, charge etc., that can take an indefinite number of determinate values. Typically, laws refer to properties – such as mass, distance, charge etc. – that can take different values, whether these are discrete or form a *continuum*.

For Aristotle, Armstrong and most metaphysicians the existence of a determinable property ontologically depends on the instantiation at some time of at least one of its values. This is not the case for Ellis. In his view, the general property of mass is *ontologically prior* to the property of having a mass of a particular value. This is because, according to him, ontological dependence must have the same direction as logical implication: A ontologically depends on B if and only if A implies B, but B does not imply A. (Ellis 2012, p. 64) For example, the existence of a water molecule logically entails the existence of an oxygen atom and two hydrogen atoms. Yet the existence of these atoms does not imply the existence of a water molecule since hydrogen and oxygen atoms can exist independently of their union in a water molecule. Therefore, the existence of the water molecule depends ontologically on the existence of the atoms that constitute it, and not the other way around. What is described by the consequent of the above implication, namely the constituents of a molecule, are ontologically prior with respect to the latter. This corresponds to the intuition according to which a composite entity is ontologically dependent on its components, especially microphysical ones.

The instantiation of a particular value of mass implies the instantiation of the general property of mass. But not conversely, because the instantiation of the generic property of mass implies that one of its values is instantiated, but not any specific one. Thus, according to Ellis, the general property of mass is ontologically prior to the instantiation of any particular value of mass. "The more general property kinds (quantities and other determinables) are ontologically more fundamental than the more specific." (Ellis 2012, p. 65).[11]

According to Ellis, every *universal property* is a *natural kind of tropes.* (Ellis 2002, p. 57) A trope is a singular instantiation of a specific property; or, in other words, a trope is the singular fact or state of affairs that exists when a universal property is instantiated. There are many objects that are spherical, but the sphericity of an object *a* is distinct from the sphericity of an object *b*, even when the two sphericities have exactly the same radius. A trope, unlike a universal, or a quasi-universal, is not repeatable, even though tropes may look exactly alike. Ellis views a property as the *kind* of all its singular instantiations, but is not the kind of the objects in which it is instantiated. A property is instantiated *in* objects and *by* its tropes. (Ellis 2001, pp. 97–98).

Brian Ellis's essentialist ontology is particularly rich, even plethoric. In addition to dispositional and categorical properties, and the individual entities in which they are instantiated, Ellis' ontology includes natural kinds which enjoy a *sui generis* reality and have essences, i.e. sets of necessary and sufficient properties which characterise members of a specific natural kind. Moreover, as we saw, natural kinds are divided into three main classes: substances, properties and processes. If a natural kind belongs to a distinct *sui generis* ontological category, it is irreducible to the class of its members and also to its essence, that is, the set of the essential properties of the members of its associated class. This is the core of the "hypothesis of essentialism". A natural kind is not reducible to its instantiations because their number may vary while the natural kind does not change. It is therefore impossible to identify a natural kind with the set of entities that instantiate it, i.e., with its extension. A natural kind is a substantive universal, a quasi-universal which is instantiated by the members of its associated extension. Being an electron is not a classical universal property instantiated in individual electrons. If this is so, being an electron is not reducible to its essence namely the set of essential properties instantiated in electrons, that is, having a characteristic mass, charge, spin etc.

Natural kinds, together with the distinction between essential and accidental properties, are the basis of the new essentialism ontology. Without this, no essentialist metaphysics can stand. However, like most empiricists, I do not think that the *sui generis* existence of natural kinds and the associated distinction between essential and accidental properties withstand strong objections, as I will endeavour to show in the next section.

[11] Unlike Ellis, I deem specific values of generic properties ontologically more fundamental (see Sect. 5.7 below).

5.2 Two Objections to Ellis' Essentialism

While Ellis' dispositional essentialism has the appreciable merit of highlighting the primordial role of causal powers in a metaphysics of nature, it also raises serious difficulties. In this section, I intend to focus on two important ones. The first hinges on Ellis' commitment to an ontology of the world grounded on natural kinds and their essences. The second difficulty regards the quiddity attributed by Ellis to categorical properties, a quiddity that threatens the metaphysical necessity of laws.

5.2.1 The Ontological Status of Natural Kinds

Natural kinds present a twofold interest in natural sciences. First, they allow us to organise nature into hierarchies and thus provide a structured overview of the domain of some specific sciences, such as physics and chemistry (but not in some areas of biology and in social sciences). Second, they play an important role in inductive inferences. If entities belong to the same natural kind, they will always (or with some probability) manifest a similar behaviour. Members of natural kinds, such as gases, electrons, gold or water samples etc. do indeed show regular behaviours described by laws, which enable us to make successful predictions.

Moreover, although the terms "essence" and "natural kind" rarely appears in the vocabulary of scientists, it is true that a significant part of scientific research, in some areas of natural sciences at least, aims at discovering the properties which are necessary and sufficient – and called "essential" – for making an entity that instantiates them a member of a specific natural kind. In fundamental physics and chemistry, such properties are microphysical properties such as the atomic number of gold (an element that contains 79 protons) or the chemical composition of water (H_2O). The knowledge of these properties and their relations with other properties described by laws makes it possible to predict and also explain the behaviour of the members of a kind.

However, in other scientific disciplines the search for essential properties of kinds is extremely difficult and even irrelevant. In evolutionary biology, the identification of the essential properties of members of the same species remains a subject of debates. For Ellis, a biological species is not a natural kind as we saw above, but a cluster of similar natural kinds. As he says, "there is no set of standing intrinsic properties or structures that would distinguish the members of these [cluster] kinds from those of all other kinds". (Ellis 2001, p.32) In social sciences, such as demography, sociology, economics etc. Ellis acknowledges that natural kinds and essences do not play a role. (This is also the reason that prediction and the presence of laws in those disciplines remains a matter of debate.)

To save essentialism, Ellis suggests that the values which give rise to our social systems originate in the biological human nature as produced by the evolution of species. (2002, p. 147) The ontology of the world would then eventually be grounded on natural kinds of elementary particles, and then atoms and chemical compounds, such as H_2O and DNA. Such claim is in line with Ellis' adoption of the principle of physicalism according to which "every world-object is necessarily a physical object: an object that has energy, and is capable of interacting causally with other physical objects." (2002, p. 86) Since it is highly dubious (or at least very controversial) that human values, legal systems, mental states etc. are physical objects, it is hard to defend the ontology of the world exclusively consists in natural kinds identified by their essential properties.

What is more, Ellis' contention that natural kinds enjoy a *sui generis* reality as a special kind of properties is hard to defend, even in fundamental physics, as I will presently show. Ellis offers the following argument in favour of the *sui generis* existence of natural kinds, an argument he illustrates with the example of the electron. Suppose that being an electron is a classical or ordinary universal instantiated in things. Then we have to assume that there is a something that has the property of being an electron. Since only an electron can be an electron (whereas many different things can instantiate an electric charge for example), the something cannot be identified independently by some other intrinsic property. We have then to postulate the existence of a neutral thing or a bare particular in which the property of being an electron is instantiated. (Ellis 2001, p. 73, pp. 91–92) But it is absurd to suppose that bare particulars exist, and on this last point I fully agree with Ellis.

As pointed out by Mumford (2004, pp. 112–113), such argument lacks persuasive force.[12] Ellis rightly says that only an electron can be an electron, whereas many different things can be negatively charged. However, *pace* Ellis, only negatively charged things can be negatively charged… It is true that linguistically, we say that "this is *an* electron", "this is *a* horse" etc. whereas we say that "this is negatively charged", "this is pink" etc. But linguistic usage does not have ontological purchase, as Ellis himself acknowledges.

What is more, the something in which the conjunctive property of being an electron is instantiated can be identified by some of its extrinsic properties as we saw in Chap. 3. Even if Ellis correctly claims that all intrinsic properties of electrons are essential to them, electrons are unambiguously identified as the entities that cause the luminous phenomena in discharge tubes, as we have seen. If one denies the possibility of identifying electrons by means of external properties, the proposition <all electrons have a negative charge of specific value> would be analytic since the term "electron" would be defined as "something that has negative charge of a specific value". Of course, if electrons are the only particles with that specific charge, then the assertion that electrons also have mass of specific value is not analytic. As we have seen above, my point is that detectable entities such as electrons must be first identified by means of external properties before we have good reasons to attribute intrinsic properties to them.

[12]Notice however that there is some mixing up between in and by instantiations in Mumford's discussion.

As a reductivist with respect to natural kinds I oppose the *sui generis* reality of natural kinds: a natural kind is just a set of entities that instantiate a conjunction of natural properties and not a substantive universal or quasi-universal. Being an electron is tantamount to instantiating the properties of having a negative charge, mass etc. of specific values, that is, a conjunctive property which is not endowed with a substantive reality on top of the existence of its conjuncts.

Nothing prevents us to decide that the set of pink negatively charged objects form a natural kind called "pinkneg". For something to be a pinkneg it must have the essential properties of being pink and negatively charged. Granted, the proposition <all pinknegs are pink and negatively charged> would be necessarily true, but only analytically so. It would not be a law of nature. Besides, such a natural kind would perhaps not be very useful for making predictions, but such concern is a pragmatic – not ontological – issue.

It is obvious that a same set of natural properties can be instantiated in a multiplicity of particular things that can be taken to constitute a natural kind. These properties form a conjunctive property the reality of which does not add anything with respect to the existence of its conjuncts. We cannot create new properties by logical operations. Ellis is certainly right on this point. Armstrong also denies that a joint property exists "in addition" to the reality of its conjuncts, and *a fortiori* that it enjoys a special type of existence. The reality of properties A and B does not imply the distinct – in surplus, so to speak – existence of a joint property A and B (Armstrong 1978, vol. II p. 30). What is more, since only individual properties, and not their conjunction as such, are observed, there is no reason to attribute to a conjunctive property an ontological weight that would add to the reality of the individual observed properties. I completely agree with Armstrong when he says that "it is at least perfectly possible to give an account of "substantive universals" purely in terms of properties." (Armstrong 1978, vol. II p. 67) It is clear that Armstrong embraces a reductionist position with respect to natural kinds.[13]

Natural kinds do exist, but I contend that they are merely sets of entities that instantiate a specific set of observed associations of properties that are necessary and sufficient for an entity to be a member of a definite kind. These properties can be called "essential" if you wish. Other properties which are not essential and might be not shared by all members of a kind can be called "accidental". However, such distinction is devoid of ontological import.

Ellis objects that the position of the reductionist is unable to *explain* the connection, which for him is necessary, between the internal properties which determine the essence of a specific natural kind.

> (...) [there are] conjunctions of properties that require some explanation better than their accidental conjoinability. (Ellis 2001, p. 97)

In the first place, it is far from obvious that the regular associations of the internal properties of what we call "electron" are necessary. But let us grant that the

[13] See Armstrong (1978, vol. II pp. 62–63, 1997, pp. 67–68).

association of properties such as specific values of negative charge and mass, as well as ½ spin (that characterize electrons) are necessary. How could the existence of a substantive universal such as being an electron *explain* that the association of these properties in its members is *necessary*? Finally, even if natural kinds did enjoy explanatory power, this would not be sufficient to justify belief in their existence.

It is true that some groupings of properties, such as the constant association of some observed superficial properties of water, can be scientifically explained by the properties of water molecules, namely their chemical composition.[14] But if it is possible to give such an explanation, at least partial, it is because laws, inductively verified and also necessary, link the properties of water molecules to certain ordinarily observed properties of water. This scientific explanation does not prove the *sui generis* existence of a natural kind. It only shows that superficial properties of water are necessarily associated with chemical properties, provided that the laws involved are necessary (and I will show later that this necessity is based on powers, without having to resort to natural kinds). We have discovered that water, previously identified by its immediately observed properties, is composed of molecules that contain two hydrogen atoms and one oxygen atom. If a substance with specific superficial properties is necessarily composed of this type of molecules, it is by virtue of necessary laws that describe how the chemical composition of water is responsible for its superficial properties. Then, it is correct to state that water is (mostly) composed of identical H_2O molecules. Given that this statement has empirical content, it is certainly not analytic. Besides, it is necessary because it is a consequence of necessary laws, not because it describes essential properties of water. We are then entitled to claim that the proposition <water is mostly H_2O > is a *bona fide* law of nature, i.e. a proposition with *de re* necessity.

Regarding electrons, first identified as the causes of specific luminous phenomena, we have good reasons to believe that its internal properties of negative charge and mass are *necessarily* conjoined since they are attributed to electrons on the basis of necessary causal, nomological connections. If specific effects are observed in discharge tubes, then the laws of mechanics and electromagnetism dictate that these effects are due to the presence of particles whose ratio of charge to mass takes on a precise value. This necessary combination of charge and mass depends on the necessary lawful connection between the cause (detected properties) and effect (directly observed luminous properties), in the same way as the chemical composition of water necessary causes the presence of its superficial properties. In this way, we have reason to believe in the necessity of the association of some detected properties of electrons, water etc. which rests on necessary laws. We need not to appeal to the existence of a substantive property whose explanatory force is, to say the least, obscure.

[14]Here I part company with Locke according to whom we would have discovered the *real essence* of water, namely its chemical composition (unknown to Locke of course), while the ordinary properties that allow us to recognise the presence of water constitute its *nominal essence*.

We are then in a position to conclude that the necessity of some associations of properties – as it is the case for some properties jointly instantiated in electrons and water – is not grounded on the *sui generis* existence of natural kinds (but on powers, as I will show below). Postulating natural kinds and essences that would allegedly explain the necessary conjunction of some properties is either an obscure explanationist (thus unjustified) move or purely definitional and devoid of ontological import.

What is important for the realist is that certain properties – powers in particular – are real. It matters little whether these properties are allegedly essential or accidental. As many empiricist philosophers have pointed out, our classifications or hierarchies are guided by observations and pragmatic considerations, among which the main one is to be able to make inductive inferences on the basis of observed similarities (Quine 1969, pp. 115–138). Neither the reality of properties, nor the constant associations of certain properties in particular entities, indicate that the world is organised in a *single* hierarchy. Admittedly, some properties are more general than others, in the sense that the extensions of some properties contain subsets of particulars with more specific properties. The property of being charged is more general than that of being negatively or positively charged. The set of negatively charged entities is included in the set of charged entities. Nothing precludes taking the set of negatively charged entities as a subspecies of the kind of charged entities. Nor does anything prevent us from considering that electrons, as they are defined, belong to a kind that has no subspecies. Pragmatic reasons can legitimately be invoked to construct such classifications. Again, such reasons do not carry ontological force. They provide no justification for postulating the *sui generis* existence of ontological stuff such as kinds, species and subspecies in addition to the properties instantiated in particulars.

If we have identified an entity as a gas, we know how it will behave. But their practical usefulness in making predictions does not imply that natural kinds enjoy an existence independent from constant groupings of properties, nor that natural kinds exist in some transcendent world (which would be a form of Platonism). It is certainly convenient to designate particulars that carry regular associations of properties by natural kind *terms* such as "electron", "lion", "gold" etc., to avoid having to use expressions like "that which has the properties F, G, H etc." Natural kind *terms* are convenient *abbreviations* for such expressions. My position is reductivist, but it is not eliminativist. Ontologically reducing natural kinds to real properties or, equivalently to classes of entities in which some identical properties are instantiated, certainly does not oblige us to eliminate natural kind terms from our language. These terms typically appear in statements like "this is *an* electron" or "this is *some* gold", and which have the following predicative form: this is *a* G or this is *some* G – and not: this is P. The semantics of these statements is distinct from those of predicative statements like "this is red", or generally "this is P".[15]

For sure, the semantics of natural kind terms is different from that of predicates. But, as Putnam (1970) and others have stressed, no ontological conclusions can be

[15]For an overview of the debates on the semantics of natural kind terms, see Bird and Tobin (2023).

legitimately drawn from semantics. To repeat, members of natural kinds are just particulars which instantiate some identical or similar properties in association. These properties and their nomological connections to other properties are sufficient to warrant predictive success. There is no need to promote natural kinds to special ontological status to explain such success as Ellis and others claim.[16] If an electron is attracted by positive charges, it is not because it belongs to the natural kind electron, but because it has the natural power of being attracted by charges of the opposite sign.

Pace Ellis, I do not think that laws of nature are explications of the essential properties of natural kinds and that any account of laws must begin with an analysis of natural kinds. Natural kinds and their supposed essences are not needed to ontologically ground the necessity of laws and to provide truthmakers of them. Natural powers, which are classical universals instantiated in things, are quite sufficient for that purpose, as I will try to show below. Scientific laws describe regularities that necessarily occur in certain circumstances because of the intrinsic powers possessed by particulars such as individual gases, massive bodies, electrons etc. There is no need to postulate some idiosyncratic existence of natural kinds to ground the *de re* necessity of laws: universal powers alone are perfectly able to serve that purpose. Moreover, since I do not believe in substantive natural kinds, I need not distinguish between "instantiation *in*" and "instantiation *by*" as Ellis does. Henceforth, I will thus stop distinguishing between instantiation in and instantiation by, except when it is required to avoid confusion.

5.2.2 Categorical Properties and the Necessity of Laws

The second objection to Ellis' essentialism relates to the metaphysical necessity of laws. In order to formulate this objection, it is necessary to revisit the tricky distinction between powers and categorical properties. Categorical properties are "properties whose identities depend not on what they do, but on what they are." (Ellis 2012, p. 19) Categorical properties are passive, whereas powers are dynamic and active. Yet it is not so simple to draw a clear distinction between active and passive properties. All the more so since Ellis endows categorical properties with a *causal role*. Indeed, the properties of position and distance, which are paradigmatic examples of categorical properties, are mentioned in many laws, including causal laws.[17]

---

[16]Predictive success is also invoked to defend the existence of natural kinds by Lowe (2002), Bird (2007) and Chakravartty (2007), among others.

[17]I defined causality by means of causal generalisations or laws, namely those that contain a time derivative (see Chap. 1). My view on causality differs from that advocated by Ellis for whom "the mechanism of causality appears to be simply that of energy transfer." (Ellis 2012, p. 12) Ellis' conception of causality echoes the one advocated by Salmon (1984) and Dowe (2000).

For Ellis, spatiotemporal positions are "properties of *instances* of the causal powers". (Ellis 2012, p. 18) Which means that spatiotemporal properties are properties of instantiated causal powers, i.e. *tropes*, but without being causal powers themselves. Despite this, spatiotemporal properties, like locations, are "causal factors":

> Being a factor in determining an outcome is not the same as being a cause, or having the causal power to achieve this outcome (. . .). Living a long way from Sidney is a factor in determining whether or not I can walk there. But it is not a physical causal power, or anything that has such a power. (Ellis 2012, p. 20)

What holds for positions and distances also holds for all categorical properties, such as geometric shapes, structural properties and so on. Given that they are not causal powers, categorical properties are passive. According to Ellis, they also are *quiddities* (and thus the objections against quiddities mentioned in the previous chapter apply). Being quiddities, their causal roles are not fixed; they can vary in different possible worlds. How then can Ellis claim, without contradicting himself, that the laws are true in all possible worlds and at the same time defend that some of the properties mentioned in them – the categorical properties – can play a causal role that varies in function of the world in which they are instantiated?

To this Ellis replies that the causal role fulfilled by categorical properties is only *accidental*. Granted, for Ellis, the necessity of laws rests on essential causal powers which are not quiddities since they perform the same causal role, i.e. give rise to the same manifestations, in all possible worlds. But these manifestations are characterised, at least in part, by categorical properties which must therefore be identical across all possible worlds. Otherwise causal powers would vary across possible worlds. Hence, categorical properties cannot be quiddities if powers are not.

Let us look more closely at this through an example. For Ellis, charge is an essential property of many natural kinds (electrons, protons etc.). Charge is a causal power that confers on the entities in which it is instantiated the possibility of manifesting a behaviour described by a law. In our world, all charged bodies obey Newton's second law with the force given by Coulomb's formula. Any charged body has the power to accelerate in the presence of another charged body, in inverse proportion to the square of the distance between them. Suppose now that in another possible world distance (and then acceleration) fulfils a causal *role* distinct from the one it plays in our world. It is difficult to understand how, in these conditions, Newton's second law coupled with Coulomb's formula, which is true in our world, could also be true in this other possible world. This law would describe other manifestations corresponding to other powers in some other worlds. So either laws are not true in all possible worlds, or distances are identical in all possible worlds. But both of these options are resolutely rejected by Ellis.

5.3 Powers

Before I present the dualist metaphysics of laws which I favour in a synthetic way, some further analysis of the crucial notion of power is in order. In line with the Aristotelian tradition, for which the observed regularities are ontologically grounded on internal principles of action possessed by substances or entities, Rom Harré and Edwin Madden argued, as early as 1975, in favour of the existence of *intrinsic and irreducible powers.*[18]

According to this neo-Aristotelian view, all existing entities instantiate internal powers or potentialities. If an entity instantiates a power P, it has the *real possibility* of manifesting a process (i.e. a series of properties successively instantiated over time) or persistence in a state (characterised by the instantiation of stable properties). Furthermore, the manifestation M (which need not be actually observed by someone) will *necessarily* happen provided that the appropriate circumstances are realised. These circumstances include a triggering event or a stimulus (or various stimuli) S, such as immersing a sugar cube in water, and an environment E that is favourable to the exercise (or manifestation) of a power by the particular which instantiates it. The (complex) properties E, S and M are categorical properties. In what follows, I will limit myself to natural dispositional properties, which I have called "natural powers" or simply "powers". Powers, denoted by "P_M", are a subset of the set of dispositions.

For example, when ingested by cyclists a quantity of amphetamines (S) has the power P to make them to manifest higher performances (M). A cyclist will necessarily manifest this power under *normal* circumstances, that is to say, an environment E which includes the conditions *ceteris paribus* (but not the triggering circumstances or stimuli). These conditions depend on the *context* in which a regularity described by a law usually occurs. Such context can be our familiar surroundings, a specific experimental setting or some natural environment (such as the interior of the human body), which need not be explicitly described. Neither is it necessary to mention all the possible interferences that would prevent the regularity to take place, all the more that exhaustively listing these is unattainable in practice. A particular which possesses the power P_M and which is submitted to an appropriate stimulus S will necessarily instantiate the corresponding manifestation M provided the relevant conditions are identical or sufficiently close to the *normal* conditions under which other similar particulars have regularly manifested the properties M. Such regular manifestations are described by what I have called "nomological formulas".[19]

If such stable normal conditions did not obtain, we would not observe any regularities (nor could we observe them because our existence and biological

---

[18]With variations, a metaphysics of powers has been adopted and developed by, among others, Sidney Shoemaker (1980), Nancy Cartwright (1989), Brian Ellis and Caroline Lierse (1994), Ellis (2001, 2002), John Bigelow et al. (1992), Stephen Mumford (1998, 2004), Mauro Dorato (2005), Alexander Bird (2007), Anjan Chakravartty (2007), Ghins (2007, 2014), Jonathan Jacobs (2010, 2017), Max Kistler (2006) and Barbara Vetter (2015).

[19]To recall what I said in Chap. 1, nomological formulas, such as $PV=KT$ or $F = ma$, are distinct from the complete formulation of a scientific law. The latter refers (most often implicitly) to the *ceteris paribus* conditions and mentions the entities for which the laws hold true, such as gases or massive bodies. In disciplines that are not mathematized, statements that describe lawful relations between properties can also considered to be nomological formulas.

constitution depend on them). Such conditions (also called "provisos") do not have to be explicit.[20] Mentioning provisos in the statement of a counterfactual or a law does not protect them from possible falsification. If some anomalies occur, scientists typically inspect the environment to detect the presence of causes susceptible of producing them. Also, they check whether the particulars involved are relevantly similar to the ones which have previously manifested the regularity. If, after careful investigation, anomalous events cannot be traced back to unusual external circumstances or misidentification of the particulars involved, then the law can be considered falsified and must be modified. Yet and this is what the realist of powers claims, if the normal (*ceteris paribus*) circumstances obtain, a particular *a* which instantiates a power P_M will *necessarily* manifest the set of properties M if the appropriate triggering circumstances S are in place.[21]

Many authors, such as Martin (1994), have objected that the instantiation of a power does not imply the corresponding manifestation even if the triggering circumstances and the normal context are in place. Antidotes or contrarians (finks), as they are called in the literature, may prevent the expected manifestation from appearing. Look for example at opium, which has the power to induce sleep. If a person who took opium also ingested an exciting substance, the latter acts as an *antidote* (or *mask,* according to the terminology favoured by Johnston (1992) and Vetter (2015)) which counterbalances the action of opium. The power of opium to induce drowsiness remains present and is also exercised, but is occulted by the counterbalancing action of the antidote.

A *contrarian* or a *fink*, on the other hand, annihilates a power at the very moment the stimulus occurs. Let us illustrate this with a standard example. A copper wire has the capacity to conduct electricity: if a difference of electric potential is applied to its ends, it will normally be traversed by a current. But one cannot exclude that when a difference of potential is applied, a contrarian causes the wire to instantaneously cease to be a conductor of electricity. The contrarian annihilates the power to transmit a current. Martin (1994) imagines a wire connected to a device that instantaneously detects the application of a difference of potential. At that very moment, this device makes the wire unable to conduct electricity. Similarly, Lewis (1997) hypothesises the contrarian intervention of a sorcerer capable to predict the moment when a difference of potential is applied and then immediately makes the live wire "dead", that is, unable to be traversed by a current. Less far-fetched

[20] See for example Elder (1994, pp. 653–654) for the justification of the recourse to normal conditions even if they cannot all be made explicit. See also Ellis (2002, pp. 116–117).

[21] Strictly speaking, it is not correct to say that "powers manifest themselves", as we often read in the literature. Only particulars manifest (or exercise) powers. When a manifestation occurs, it may result in the disappearance of a power (such as solubility, when a lump of sugar dissolves) or its permanence (like the capacity to roll, when a ball rolls).

examples include food losing its healthy nutritive power when being infected by bacteria and acquiring the power of being poisonous.[22]

For a realist of powers, the possible presence of a contrarian or a fink is not a problem because in this case a given power simply ceases to be instantiated. We are therefore dealing with *another* particular. It is clear that a particular must first possess the relevant power in order to have the possibility of acquiring the corresponding manifestation. Actually, the problem raised by contrarians is epistemic. If some food has become infected unbeknownst to us, then we attribute a wrong power to it. Infected food is not similar to nutrients which have regularly manifested a healthy feeling of satiety, even if it has the appearance of being so. Then, the ingestion of food which has been unknowingly infected, is not followed by the expected manifestation. Similarly, the unknown presence of an electrofink impairs the reliability of a prediction: contrary to what was expected, no current flows in the wire which has become something else than copper.[23] At the ontological level, there is no difficulty: by annihilating a power, an electrofink transforms the initial copper wire into something else.

Antidotes or masks, on the other hand, do not change the nature of particulars. The nature of opium, which has the power to induce sleep, remains unchanged. The manifestation of drowsiness really occurs, although it remains invisible, since it is *masked* or concealed by the action of the antidote. Recall that the term "manifestation" is misleading here since this term usually refers to something which is observed. Yet the manifestation corresponding to a power can happen when unobserved and unbeknownst to us. True, the unknown presence of antidotes or masks prevents making accurate predictions. But again we are confronted with an epistemological issue. In normal circumstances and if triggered, particulars endowed with actual powers will instantiate the corresponding manifestations. The situation in the case of antidotes is analogous to, for example, the equilibrium of the two forces, gravitation and resistance, acting on my cup when it is at rest on my desk. The two forces are real and acting, even if my cup does not move.

Like Barbara Vetter (2015, p. 115), I will take powers or potentialities to be *ontologically primitive* properties. Powers ontologically ground real possibility. A potentiality endows a particular with the possibility of engaging into some specific manifestation and is thus identified by the latter only. Hence, the identification of a power does not include a description of the conditions in which its bearers will

---

[22] A summary of the discussions on this issue can be found in Bird (2007, Chap. 2). See also Mumford (1998, pp. 81ff). Like Ellis and Chakravartty, I will not get into many details on the discussion about finks and antidotes.

[23] Just as the proposition <water is H_2O > is a law, <copper conducts electricity> can be considered a law if copper is identified by its superficial properties.

instantiate its possible manifestation. In some contexts[24] the particular – necessarily – will manifest the behaviour made possible by the power, in others no.

All particular entities have several sorts or types of powers. A sort of power is characterised by a definite set of possible manifestations, which can sometimes be described by a nomological formula. Thus, a negatively charged particular, such as an electron, has the power to be attracted by a positive charge. The value of the force of attraction varies in function of the distance between the charges and the size of the positive charge. But an electron, if it is in motion, also has the possibility to be accelerated in a magnetic field. In each case, the set of manifestations is described by specific nomological formulas or laws of action, which, as Ellis says, describe kinds of processes.

For some authors,[25] there exist powers that give rise to different types of manifestations depending on the circumstances. These are known as *multi-track powers*. For example, elasticity is a power that gives its bearer the capacity to increase or decrease in volume, but also to bounce etc. It seems to me that a multi-track power can be analysed in several *simple* powers that are distinguished from each other by specific types or kinds of possible manifestations. On this, I agree with Molnar (2003, pp. 198–199) who claims that there are *simple* powers *directed towards one type or kind of manifestations*. Multi-track powers, such as elasticity, can be reduced to a set or *cluster* of simple univocal powers. Different types of manifestations will usually be triggered by different types of stimuli. Thus, a rubber ball will bounce if thrown on a hard floor, while its shape will change if it is crushed in the hand etc. Yet what distinguishes and individuates a power is the type of manifestations it makes possible. We are talking about *types* of manifestations because the ways in which a ball can bounce etc. are multiple and varied. Nevertheless, I will consider the power of bouncing as a single-track disposition or simply a disposition.

In the natural sciences, powers can generally be accurately defined and be considered simple. Each power can give rise to a single kind of manifestations that corresponds to the satisfaction of a specific functional law. Thus, a charged body has the power to accelerate in the presence of an electric field, in accordance with Newton's second law and Coulomb's mathematical expression of the electrostatic force. A charged body also has the distinct power to move in a magnetic field under the action of the Lorentz force. To each initial velocity of a charge in a magnetic field corresponds some acceleration. Yet all these accelerations belong to the type of manifestations described by the same nomological formula. Thus, I will consider the

[24] Surely, dispositions are sensitive to context. As Barbara Vetter (2015) remarks, a diamond as well as a Chinese vase have the power to break. But a diamond is not generally considered to be fragile, while a Chinese vase usually is. In other words, there is a gradability of dispositions: it is harder to break diamond than a Chinese vase... But a diamond has the real power to break (and is thus fragile) in the context of being hit by a hammer. To provide a general characterisation of degrees of powers (or potentialities, as Vetter calls them) is a daunting task. Given the wide variety of contexts on which such degrees depend, I personally doubt that such aim can be attained.

[25] See Ryle (1963, p. 114) and Bird (2007, p. 21)

power of a charge to interact with another charge according to the Coulomb formula as a single-track disposition.

According to some authors (Vetter (2015), Ioannidis et al. (2021)) functional laws raise major difficulties for a metaphysics which resorts to powers only. A first difficulty is the following. Since a functional law, such as the Coulomb law, covers an infinite number of possible quantitative manifestations, we should attribute a specific power to each value of the interaction between two charges. Such a situation leads the power monist to accept the existence of an infinity of powers instantiated by each charged particular. But I don't think the Coulomb formula is equivalent to an infinite conjunction of specific quantitative manifestations. In fact, the formula functions in a way similar to universal predicates. The predicate "red" can be instantiated by particulars having various shades of red such as amaranth, scarlet, purple, crimson, carmine, Bordeaux etc. In an analogous way, the power of green apple to mature into a red apple is a single-track disposition.

Another difficulty is that in the case of functional laws such as the Coulomb law, the triggering circumstances are part of the manifestation. In fact, to manifest a specific quantitative interaction with another charge, a charged particular has to be located at a definite distance of another charge. The manifestation described by the Coulomb formula mentions the distance between the charges. Therefore, in many cases, it is not possible to distinguish sharply the triggering circumstances or stimulus from the manifestation. I acknowledge that such a situation is irksome for those who attempt to logically formalise dispositions in terms of biconditional analyses, such as Alexander Bird (2007, pp. 24 ff) and others have proposed. Following Barbara Vetter, I defend a primitive notion of power, as said above. A power endows the particular which instantiates it with the possibility of engaging in some behaviour in certain circumstances, without having to precisely describe these circumstances and without demanding that these circumstances be prior or independent of the manifestation. When submitted to the stimulus of being immersed in water, a sugar cube dissolves. While manifesting the process of dissolution, the stimulus is still present, otherwise the sugar cube would stop dissolving. When the process of dissolution is achieved, the dissolved sugar is still in water.

A third, and in my opinion major, difficulty concerns the combination of powers. When two charges are the same sign repel each other, they both exercise their own power of manifesting a behaviour described by the same Coulomb formula. In such a situation, Ioannidis et al. wonder "how the various contributions of the powers will be combined, so that Coulomb's law will be satisfied?" (Ioannidis et al. 2021) The problem can be generalised. If a sugar cube has the power of dissolving in water, water also has the power of dissolving sugar. Other examples abound. I confess that I don't have a satisfactory solution to this problem within a monistic ontology of powers.

However, I suspect this is a pseudo problem arising from the anthropological connotations associated with powers. As Ioannidis et al. (*Ibid.*) ask "how does the concrete charge happen to be sensitive to the right kind of information available in the inventory of possible values in such a way that Coulomb's law is satisfied?" It seems to me that is enough that the manifestation is accurately characterised by the

Coulomb formula, in which the various mentioned properties are real. Is this a satisfactory *dissolution* of the problem raised by Ioannidis et al.? Probably not. But it is certainly not a satisfactory *explanation* of the problem they raised. On the other hand, the explanation they propose by resorting to real laws which govern the interaction between charges is not immune of problems either. Let me only mention the mysterious nature of the governing in play.

5.4 The Relation Between Powers and Laws

Powers provide "natural necessity and possibility, and are fit to be truthmakers of modal truths", as Mumford (2004, p. 170) puts it. This assertion needs to be clarified. We are grappling here with a vexing issue which is still fiercely debated. In what follows, I will draw much inspiration from Stephen Mumford without pretending to offer a definitive solution to the ontological grounding of the necessity of nomological propositions.

Laws, as formulated in scientific theories, describe regularities and are purely extensional (non-modal). For the realist of powers, a nomological formula describe *possible manifestations* by the particulars that instantiate a specific power. These manifestations are not necessary; they are merely possible. Nevertheless, the bearer of a power will necessarily instantiate the corresponding manifestation if the appropriate stimuli S and the normal environment E are in place. There is thus a link between powers and counterfactual conditionals. The possession of a power by a particular implies that it will necessarily manifest a specific behaviour when submitted to the triggering stimuli in normal circumstances. Hence, the possession of a power by a particular is *part* of the truthmaker of counterfactual propositions such as: <in normal circumstances, if the triggering stimuli occurred, a given particular would manifest a specific behaviour>. For example, <in normal circumstances, if a sugar cube were immersed in a glass of water, it would dissolve>. A power is only part of the truthmaker of a counterfactual because its truth also depends on the normal circumstances and appropriate stimuli. The counterfactual <if a sugar cube were put on my dry desk, it would dissolve> is obviously false. The mentioned activating circumstances are not the normal circumstances in which the manifestation occurs even though the power of solubility is actually instantiated in sugar cubes.

To say that a particular is soluble *means* that it has the possibility of dissolving under certain circumstances. The proposition <a particular is soluble if and only if it has the possibility of dissolving> is a necessary truth, but it is analytic since it depends on the very meaning of the term "soluble". This proposition, just as logical laws, is true in all logically possible worlds, as we usually say. Thus, the enemies of powers are quite right to claim that propositions such as <all soluble bodies have the possibility of dissolving>, <all sleep-inducing substances have the possibility of making one sleep> are analytic statements, just like the sempiternal <all bachelors are unmarried>, <all sons have a mother> etc. The truth of these propositions is necessary *de dicto*, i.e. it rests exclusively on the meanings of the terms used and not on any reality. This is all very well, but as I insisted the necessity the realist is interested in is real – *de re* necessity.

The issue at stake here hinges on the *real necessity of the connection* between distinct properties, namely a power and its corresponding manifestation. Notice first that it would be hard to deny that the meanings of "power" and "manifestation" are distinct. Powers and their manifestation can be instantiated at different times, and sometimes they have to. A sugar cube cannot be both soluble and dissolved. Ontologically, the state of being dissolved necessarily depends on the prior instantiation of a power in the sugar cube. For the friends of powers, the connection between a power and its manifestation is a real causal necessary link. Yet such causality is distinct from Millian causality: it is *metaphysical*. In his critique of causality, Hume challenged the existence of such a necessary link in nature since it transcends any possible observation. But he certainly did *not* say that the claim that there is a necessary connection between distinct facts, namely the instantiation of a power on the one hand and a manifestation on the other, is analytic and empty. As Stephen Mumford points out "(...) it would be absurd to say that *the cause of F* could not cause *F* because it means *the cause of F*. (...) *the cause of F* and *F* denote distinct properties and likewise for a disposition and its manifestation." (2004, p. 169).

No doubt, the proposition <the cause of manifestation M causes M > is a necessary analytic statement. Yet this proposition does not say that there is a real necessary connection between what is actually the cause of M, namely the power P, and M (in normal circumstances). In contrast realists hold that we are dealing here with a *de re* necessity, a claim which was adamantly opposed by Hume.

Powers have causal efficacy. This is why they are called "causal powers". The metaphysical causality discussed here should not be confused with the causality involved in a scientific causal law, as defined in the first chapter. First, as we saw, scientific causal laws describe changes, whereas the laws that describe a state (like the ideal gas law) are not causal scientific laws. Second, scientific causality is not associated with any necessity. Scientific laws, in the way they have been defined, are purely extensional. On the other hand, if a power has causal efficacy, a particular can instantiate a manifestation only if it possesses or has possessed the power corresponding to that manifestation. There is no manifestation without power. This also holds for scientific laws which are not causal, such as the ideal gas law. In this case, we can say that gases have the metaphysical causal power to stay in a state of equilibrium.

If powers exist (the reasons for belief in powers is a difficult epistemological issue which will be tackled below), these are the partial truthmakers of modal truths such as <if a sugar cube were immersed in water, then it would dissolve (in normal contexts)>. The necessary truth of counterfactuals rests on powers. As a matter of fact, a sugar cube, defined and identified by observed categorical properties, also has the power to dissolve. Consequently, if all sugar cubes possess the power to dissolve, this will give rise to a regularity described by a scientific law: <all sugar cubes dissolve when immersed in water (in normal contexts)>. The scientific laws which describe regularities not only are universally true but also are necessarily so. The fact that a particular possesses a power compels it, as a matter of metaphysical necessity, to behave in a certain way. In this sense, powers are *regularity enforcers*, as Stathis Psillos put it. If what has been said above is correct, powers ground the necessity of lawful propositions as well as the truth of counterfactual conditionals.

According to the version of dispositional realism that I advocate, the particulars that populate the world have powers that are *intrinsic to* them. It is true that in science these powers are capacities for relations, since nomological formulas describe relations between properties. A massive body has the capacity to attract another massive body, and attraction is a relation between two bodies. However, it is the manifestation – body a attracts body b (and vice versa) – that is a relation, whereas the power to attract is a monadic property.[26] Such powers are not properties of properties, but properties intrinsic to the particulars which carry them.[27] Surely, a particular will only exercise this power if certain relations with its environment are realised (triggering circumstances etc.), but in itself a power is a monadic property. Moreover, powers are ontologically on a par with categorical properties since they are real, actualised, properties. For example, a negatively charged body has the real power to accelerate a positive charge in accordance with Newton's law and Coulomb's formula. This power is distinct from being charged, which is a categorical property. Remember that I take being charged to be an observable property since once we have grasped the concept of charge, we can observe for example *that* charges of the same sign are present by looking at the repulsion of gold foils in an electroscope. Often the word "charge" is used to refer to the disposition to attract or repel, as Ellis and others do. Such confusion is devoid of ontological purchase. For the sake of clarity, expressions such as "charge" and "power to attract or repel other charges" should be used. The former refers to a categorical property, whereas the latter refers to a dispositional property.

---

[26]The possibility of a universe that contains only one body does not prove that powers are relational – not intrinsic – properties. It is true that in such a universe, a solitary body would never manifest a supposed power to attract another. But intrinsic monadic powers exist independently of their manifestations, and thus irrespective of our having epistemic access to them or not. In our universe many powers are never manifested, due to contingent circumstances. This does not prevent powers from existing.

[27]Since powers are not properties of properties but properties of particulars, this makes it possible to counter some objections according to which defending the existence of powers leads to infinite regress (see for example Psillos 2006). The regress objection rests on the assumption that a power has itself a power. It goes like this. Let us assume that power F (fragility, for example) has the power G to manifest *itself*. This power G is distinct from F. G is directed towards a manifestation, namely the manifestation of F. Again, G has the power to manifest itself etc. and we are dragged into an endless regress. Powers are not endowed with powers; they *are* powers. Here again language misleads us when we say, for example, that fragility, when not manifested, has the power to manifest. What has the power to manifest, that is to say, to acquire certain properties, is the particular bearer (the vase) of the power, as I stressed. A manifestation is an instantiated process (the breaking of the vase) or a final state (the broken vase).

The identity of a power P_M is fixed by the type of manifestations it makes possible for the particular which bears it. If the appropriate triggering circumstances S and normal environment E are in place, the manifestation will necessarily occur. But properties S and E are obviously not all necessarily instantiated. Thus, the reality of a power clearly does *not* depend on the – contingent – instantiation of the properties S and E, nor on the occurrence of its corresponding manifestation. Powers are thus real, occurring, actual properties, in the same way as the categorical properties are, provided they are instantiated. If so, the distinction between actual and non-actual properties is inadequate to differentiate categorical properties from powers, and also from dispositional properties in general. It thus makes no sense of speaking of the actualisation of a power. On the other hand, it is quite appropriate to speak of a particular exercising or manifesting a power, that is, instantiating or actualising its corresponding manifestation.

I have emphasised that we must distinguish between the causal relation between events that are instantiations of categorical properties, such as the succession of planetary positions in time, and the metaphysical causality between a power and its manifestation. The necessity of the causal link between events described by scientific laws is ultimately grounded on the existence of a power, i.e. as a matter of metaphysical necessity. In this sense, scientific laws are also laws of nature.

According to this neo-Aristotelian approach, the occurrence of regularities described by scientific laws is no longer a radically contingent fact, as the Humeans maintain. Surely, such occurrence depends on the instantiations of powers, triggering events and normal circumstances. Admittedly, such instantiations are contingent. But when they obtain, regularities necessarily occur. In this way, the reality of powers entails a *de re conditional* necessity in nature. As Chakravartty says, "In a given world, such as the actual one in which we live, there is *ex hypothesi* a substantial reason for the uniformity of these [causal] processes and relations." (2007, p. 127) Supposing the existence of powers provides this reason.[28] This in no way excludes the possibility of worlds in which other powers would give rise to other uniformities.

The necessity of regularities in nature is thus conditioned upon the contingent existence of powers, and properties S and E. To distinguish this *de re* necessity from logical necessity, most authors call it natural or metaphysical. Mumford and Chakravartty prefer to call it natural because the instantiation of a specific power is not necessary but contingent. While I agree with this view, I nevertheless prefer to speak of metaphysical necessity because powers are not immediately accessible to us in perceptual experience. However, as we saw, natural powers are not known and identified *a priori* but discovered *a posteriori* by observing or detecting their manifestations in the course of scientific investigation.

With the above clarifications in mind, we are now in a position to address a classical objection raised against a metaphysics of powers, namely the so-called

[28] True, the existence of powers – metaphysically – explain why there are regularities. But this does not provide a reason to believe in their existence. Such reasons will be discussed below.

"dormitive virtue" objection, which has led many philosophers to reject, sometimes summarily, and not without a touch of irony, a metaphysics whose Aristotelian heritage is all too obvious. Hasn't such a metaphysics been amply refuted, both scientifically and philosophically, a long time ago? As the opponents to Aristotelian metaphysics like to remind us, the attempt to explain the capacity of opium to induce sleep by resorting to a putative internal *virtus dormitiva* was ridiculed by Molière in *Le malade imaginaire*.[29]

However, if we are presented with a sample of opium, or any other type of substance, and told that it is soporific, this gives us valuable information about what we can expect if we ingest it. What is at stake is not to provide an "explanation" of the property of inducing sleep by the presence of a "dormitive virtue", which would certainly be not an explanation at all, but to explain why someone is drowsy by appealing to an *internal* property of opium and not to external causes. Several causes of drowsiness are indeed possible. The ingestion of a soporific substance is one of them, as well as fatigue, boredom etc. Yet to explain why opium makes you sleep because it has a dormitive virtue or a soporific power is obviously not an explanation, since it is based on the very meaning of the word "soporific". To be soporific is, by definition, a power to induce sleep in appropriate circumstances. In other words, such power is a dormitive virtue. <All soporific substances have a dormitive virtue> is an analytic proposition that provides no explanation.

On the contrary, to state that opium, identified by some observed categorical properties, is soporific is not making an analytic statement. Also, claiming that there is a real necessary connection between the dispositional property of being soporific and the categorical property of being sleepy is not to state an analytic proposition and does provide a "substantial reason" for the occurrence of some regularities.

I concede that calling this power "*virtus dormitiva*" does not shed a particularly bright explanatory light on the mode of action of this power. This kind of verbosity reveals more a desire to impress an audience with a (limited) knowledge of Latin, rather than a serious attempt to provide an acceptable explanation. It is this pedantic attitude, more than Aristotle's physics, that was perhaps the real target of Molière's irony.

Postulating the existence of an internal power in opium does explain why a person who takes it becomes drowsy (in normal contexts). As we have seen above, instantiations of the same power impose the occurrence of a regularity. In postulating the existence of a power, we do not limit ourselves to describing again a regular process with new words. Instead, we bring in an additional element, namely that this regular behaviour is (metaphysically) caused by a power rooted in particular entities.

It should also be emphasized that a causal explanation which mobilises powers is not scientific but genuinely philosophical and metaphysical since scientific theories

[29] Here is the original text: « Bachelierus: Mihi a docto Doctore domandatur causam et rationem quare opium facit dormire. A quoi respondeo: quia est in eo Virtus dormitiva cujus est natura sensus assoupire. »

describe and explain processes by means of categorical properties. Appealing to a potentiality does not answer the following question: *how* – by what *causal mechanism* – does opium induce sleep? A scientific explanation of the capacity of opium to induce sleep must be based on a mechanism and thus on the – categorical – properties of the chemical components that interact with the human organism in a way accurately described by scientific causal laws. Categorical properties are the *causal categorical basis* of powers. Yet if science tells us that molecular categorical properties must be possessed by opium for it to be soporific, this does not imply that they are sufficient! The soporific power of opium cannot be *reduced* or defined by using categorical properties only, simply because categorical properties are devoid of any modal force[30]: they are impotent to ground necessity. We thus realize that the reference to a "categorical basis" is unfortunate because the term "basis" naturally suggests that the reality of a power is grounded on such a basis. But this is not correct. Instantiated powers have a reality on their own, which doesn't need to be supported by something else.

Even if the categorical basis for a power, such as the power to make people sleep, is known, the challenge of finding an explanation for the regularities will arise at a deeper level, be it molecular or even sub-molecular. By what mechanism do some chemical substances always react in the same way with other substances? No complete and definitive scientific explanation can ever be formulated for the behaviour of a given system, since there is in principle no end point to the possible decomposition of its parts into more elementary and more fundamental elements.

Are there fundamental powers, that is, powers that lack categorical basis? I don't think we are in a position to provide a definitive answer to that question, which is epistemological in kind. True, we cannot exclude *a priori* that some powers lack a categorical basis. But it is incumbent on the scientific approach to always seek to discover a mechanism underlying their manifestations because otherwise we don't fully understand *how* manifestations are produced. However, there is no guarantee that such understanding will always be provided since there might be fundamental powers after all. Thus, we don't know, neither can we know, whether there are fundamental powers or not. Suppose now that there are no fundamental powers. In this case, an infinite number of properties, categorical and dispositional, would exist. To my knowledge, no cogent argument against the existence of an infinite number of properties is available. Besides, even if there are fundamental powers, this would not exclude the possibility of existing and infinite number of properties either.

Be it as it may, no one can impose *a priori* an endpoint to the quest for explanation in science. For this reason, appeal to powers can in no way be interpreted as an attempt to set a limit to the search for explanation in science. On the contrary, positing the existence of a power is an invitation to discover the possible categorical properties underlying its manifestation. This is why we must look again at the important role of categorical properties in the sciences.

---

[30] I don't follow philosophers who, like Stathis Psillos (2006), claim that a disposition can be reduced or defined by its categorical basis alone.

5.5 Categorical Properties

Modern science was developed in staunch opposition to Aristotelian physics: this is a well-documented historical fact. This opposition is the origin of the discredit which has affected any metaphysics of dispositional properties and powers since the seventeenth century. At the dawn of modern times, the project of mathematizing nature relied on a construal of natural entities as mere machines or mechanisms, i.e. sets of moving parts endowed with quantifiable geometrical and arithmetical properties, called *primary qualities*.[31] When instantiated, these properties stand in spatial and temporal relations described by mathematical laws. For modern thinkers, all the other properties of a system, in particular the powers to act on other systems such as our sensory apparatus, are in principle reducible to primary qualities. According to this mechanistic ontology, it is in principle possible to reduce all the properties of natural entities and their parts, as well as the relations between them, to "number, figure, and motion" (Locke 1689, Book II, §17). These are the only real properties belonging to the bodies that make up the universe. Such universe is nothing else than a gigantic mechanism whose behaviour is described by the laws of nature, or even *governed* by them if laws are taken to be real, as Descartes did. For seventeenth century thinkers, primary properties provide the foundations for building the new mathematical science of nature.

Several classical texts promote this project of mathematisation of science grounded on a mechanistic ontology. Galileo was first to proclaim the objective reality of primary qualities as opposed to qualities such as odours, tastes and sounds, which are deemed to be subjective and thus have no place in science. As early as 1623, he wrote in *Il Saggiatore*[32]:

> To excite in us tastes, odours, and sounds I believe that nothing is required in external bodies except shapes, numbers, and slow or rapid movements. I think that if ears, tongues, and noses were removed, shapes and numbers and motions would remain, but not odours or tastes or sounds. (Galileo 1957, pp. 276–277).

In this well-known citation, the words "I believe" and "I think" are obviously devoid of any argumentative force. What we have here is the expression of an act of faith, which will be abundantly justified by its subsequent fruitfulness in science. (We had to wait for Descartes for a full-fledged philosophical argument in favour of a mechanistic ontology of nature). According to Galileo, the powers of external bodies to "excite" sensations in us solely depend on their arithmetical and geometrical properties. These powers are for him reducible to the primary qualities, being the

[31] The expression "primary quality" was introduced by Robert Boyle in 1666. Before him, Galileo and Descartes had already introduced the distinction between what they called "first qualities" and "second qualities".

[32] A *saggiatore* was an "assayer" capable of verifying whether an object contained only gold, to the exclusion of any other metal. For this purpose, a high-precision balance was used. In *Il Saggiatore*, Galileo weighs arguments pro and contra Aristotelian cosmology in order to show that his arguments and observations carry more weight than those of the Aristotelians.

only ones which are objective. All other properties – the *secondary qualities*, such as colour and sound – are the result of an interaction (via corpuscles endowed with mathematical properties) between the external bodies and the sense organs, also construed as mechanisms. Secondary qualities were deemed to be subjective whereas the supposed mechanistic processes causing sensations were considered objective. Of course, no one in the seventeenth century could describe the mechanisms by which coloured bodies, for example, produce, together with light,[33] the sensations of colour. Nevertheless, Galileo thus opened the way to a research programme that was to prove extraordinarily fertile for scientific research.

Later on, Descartes took over the Galilean distinction between primary and secondary qualities and made it the basis of his mechanistic philosophy of nature. In his *Principles of Philosophy,* first published in 1644, he writes that "Therefore, all the matter in the whole universe is of one and the same kind; since all matter is identified [as such] solely by the fact that it is extended. Moreover, all the properties which we clearly perceive in it are reducible to the sole fact that it is divisible and its parts movable; and that it is therefore capable of all the dispositions which we perceive can result from the movement of its parts." (Descartes 1982, Book II § 23, p. 50) In 1674, Robert Boyle conjectured that the "solidity, taste etc. of salt may be fairly accounted for by the stiffness, sharpness, and other mechanical infections, of the minute particles whereof salts consist." (Boyle 1991, p. 149).

For Galileo, Descartes and Boyle, the primary qualities are *necessary* conditions for the action of bodies on each other, especially on our sense organs. Without these primary qualities, bodies would possess no power. It is worth pointing out that these authors often describe primary qualities as *powers* to cause sensations, such as the sensations of smell and colour. But, in fact, the primary qualities are what we now call the "categorical basis" of these powers. Colours, smells etc. are, for these philosophers, secondary qualities produced by the interactions between things and our sensory organs. It does not follow that primary qualities are powers. The supposed property for a corpuscle of having a pointed shape is not identical with the power to cause a salty sensation.

Similarly, the property of being spherical is not identical to the capacity to be observed spherical. For sure, one can only ascertain that a particular is spherical by observing it, and several perceptual modalities (sight, touch) allow us to do so. But this epistemic fact has no incidence on the ontological status – modal or not – of sphericity. The manifestation of the power to be observed spherical simply is to be observed spherical. Indeed, neither the (monadic) possibility of being observed spherical nor the (relational) property of actually being observed spherical is identical with the (monadic) property of being spherical. As such, the property of

[33] Even today, a complete explanation of the processes of colour perception is not available.

sphericity does not involve any modality, unlike the – distinct – modal property of being observable as spherical.[34] Sphericity is a primary, categorical quality.

For most philosophers, categorical properties are characterised by their *quiddity*, as we saw in the previous chapter. According to quidditism, the identity of a categorical property is primitive. Its identity is what it is, that's all: it is impossible to say more on what this identity consists in. The powers that could be associated with categorical properties are externally imposed on them by ontologically thick laws. Treading in Descartes' footsteps on this point, Armstrong, Dretske and Tooley claim that laws endow categorical properties with powers: laws make categorical properties powerful. As a consequence, a categorical property can have different causal roles in distinct possible worlds, depending on the laws that exist in these worlds. Despite of this, a categorical property enjoys a primitive transworld identity, that is to say an identity across possible worlds. Since I discussed the difficulties encountered by quidditism in the previous chapter, I will not return to them here.

Contrary to most authors, it seems plausible to me to defend that categorical properties, as well as powers, are the same in all possible worlds. Indeed, the manifestations of powers are described by relations between categorical properties. Powers are individuated by their manifestations. If powers are not quiddities, which is granted by the realists about powers, a power identified in our world remains the same in another world where it is instantiated; otherwise, it would be another power. If a specific power is the same in all worlds in which it is instantiated, so must be the categorical properties that describe their manifestations. For example, I find it difficult to argue that in some worlds spatial distances are different from what they are in the actual world, as I showed in Sect. 5.2.2 above.

But powers are not observable. Since directly observed properties exist, they must be categorical. Indeed, as we saw, observability is the mark of categoricity. All observable properties are categorical properties. Granted, very small or very large values of a categorical property like speed are not directly observable. But they are detectable. We are then in a position to claim *that a property that is directly or indirectly observable, i.e. detectable, is a categorical property.*[35] In this way, we are able to identify some categorical properties (the ones we may have reasons to believe that they exist) in a positive way, which is independent of identifying them negatively in opposition to dispositional properties.

The primary qualities of the founders of modern science can certainly be deemed to be categorical properties. Today, the project of mathematising nature has been considerably enlarged to include many other measured and calculated properties, in addition to the properties of "number, figure and motion". To the extent that such mathematized properties have been either directly observed or detected, they are *a*

[34] In our world, there is a necessary link between the properties of being spherical and of having the possibility of being observed spherical, but such necessity is grounded on laws. I will come back to this issue in due course.

[35] The converse is not true since observability is an epistemic notion. It may be that there are categorical properties that are not observable. But then we have no reason to believe in their existence.

fortiori either directly observable or detectable. This being so, restricting the class of categorical properties to geometric and arithmetic properties and to the spatial and temporal relations between them seems unduly restrictive and inadequate. In the class of categorical properties we must include all real properties and relations corresponding to the mathematical symbols used for the formulation of scientific laws (which we have reason to believe to be true), and also the observable (in the broad sense of observability) properties referred to by the terms that occur in the nomological statements of theories that are not mathematized, especially in the social sciences.

As we saw, mass is a categorical property, not a dispositional one. An electron has a mass of a precise value, as determined by using reliable measuring instruments. An electron also has the distinct dispositional property that its mass is detectable, using devices that enhance our perceptual faculties. Again, this detectability depends on the existence of powers whose manifestations are described by nomological formulas. Not all of these are known since we do not yet have a complete theory of the mechanisms of sensory perception. Of course, all this does not preclude the possible existence of other, more fundamental, categorical properties to which properties such as charge could be reduced, in the same way that heat was reduced to the kinetic energy of molecules by Boltzmann. Are there simple, absolutely irreducible properties in nature? Are there particulars that are genuine atoms, that is, indivisible?[36] We must acknowledge that experience does not consent to provide answers to these questions.

5.6 The Limits of Categorical Explanation

Explaining certainly is one of the main objectives of scientific activity. Even if the inference to the best explanation is devoid of any ontological force, it undoubtedly plays a crucial heuristic role, since it allows us to envision the hypothetical existence of entities endowed with properties for which it will sometimes be possible to empirically ascertain the presence of a causal link with directly observed properties. However, a scientific explanation of the behaviour of particulars can only be based on categorical properties. The construction of models serves this objective. A model is supposed to partially represent a real system, which, viewed within the objectivist attitude, is nothing else than a set of categorical properties organised into a structure by means of categorical relations. If we know the causal laws that describe the evolution of these properties over time, we also get a mechanistic explanation of the behaviour of the system, as detailed in the first chapter.

[36] However, it is doubtful that elementary particles can be *spatially* divided. The proton, for example, is not spatially organised into three quarks.

Take the example of optically active substances. Chemical analysis of sugars reveals that they contain six carbon atoms.[37] We observe that some sugars rotate the plane of polarisation[38] of light sometimes to the left and sometimes to the right in relation to an instrument which detects the properties of the incoming light. This makes it possible to distinguish two kinds of sugar by assigning them the dispositional property *levorotatory* (for a power to rotate the plane of polarisation to the left) and *dextrorotatory* (for a rotation to the right). Glucose is dextrorotatory while fructose is levorotatory. Distinct laws describe the different optical behaviours of glucoses and fructoses. Of course, scientists did not rest satisfied with simply assigning levorotatory or dextrorotatory powers to some substances. They endeavoured to provide a *structural categorical explanation* of these facts. Although glucose and fructose are made of the same atoms, the levorotatory fructose L is the mirror image of the dextrorotatory glucose D. By an appropriate mathematical transformation (a reflection with respect to a plane) we obtain the levorotatory glucose L from the dextrorotatory glucose D, and conversely. A L sugar can be transformed into a D sugar by reflection, just as our right hand and left hand can be. These two kinds of sugar are called *enantiomers*.

The interaction of light with L and D sugars is described by the laws of electromagnetism. Through these we get an explanation of how L and D sugars cause the deflection of polarised light in function of their distinct categorical structures. We have thus explained, in a mechanistic way, why light behaves differently depending on whether it passes through a homogeneous solution containing sugar of one or the other structure. In the same manner, one can give such an explanation for the dissolution of sugar in water by showing how the sugar molecules associate with the molecules of the solvent.

In such explanations, we have *not* reduced the power to rotate the plane of polarisation to a set of categorical properties. The causal – categorical – basis certainly explains how rotation occurs through the interaction of light and the structure of molecules. If we have good reasons to believe that the properties of the causal basis exist, we have discovered the *necessary* conditions for a power to rotate light to the left or to the right. But they are certainly not sufficient since powers are modal properties. The instantiation of the power to rotate implies the existence of its categorical basis, not the reverse. Powers do not supervene on their categorical bases. They are primitive properties, as I argued above.

The scientific explanation of the optical activity of sugars stops at this point. But we can keep asking other questions and wondering why carbon, oxygen and hydrogen atoms combine into sugar molecules. This is of course a question distinct from the origin of optical activity, but it is obviously scientifically relevant. Its answer is provided by quantum mechanics. Ultimately, we will have to resort to the properties of the components of atoms, in particular the – categorical – properties

---

[37]This is why they are called “hexoses” with chemical formula $C_6 H_{12} O_6$.

[38]The plane of polarisation is defined by the direction of propagation of the light wave and the direction of its electric field.

of electrons. When we get to the level of electrons and other elementary particles, we will be obliged to put an end – at least provisionally – to the search for a mechanistic explanation of the exercise of their powers by means of categorical properties. But nothing prevents us from also attributing powers to electrons, such as the power to behave in accordance with a particular nomological formula, even if the mechanism underlying this behaviour (if it exists) is not known at this time.

The availability of mathematical formulas is not sufficient to explain how particulars exercise a power, such as gravitation. Such formulas must also correspond to causal laws which describe the behaviour of mechanisms. Newton's formula for universal gravitation does not explain how massive bodies attract each other. As Newton famously acknowledges in his third letter to Bentley: "Gravity must be caused by an agent acting constantly according to certain laws". (Newton 1961, p. 254) However, Newton very explicitly refrains from devising "hypotheses" about such agent.[39]

Today, the explanation of gravitation mentions particles, namely gravitons, whose behaviour is described by causal laws. Recent empirical evidence for the reality of gravitons associated with gravitational waves[40] confirmed the existence of a mechanism for gravitation. But again, how do we explain the way in which these gravitons act? A classical way of explaining in the natural sciences consists of the decomposition, whether spatial or not, of an entity into a system of categorical properties structured by mathematical relations. There is no reason to pronounce, as a matter of principle, that the explanatory process has reached a final point, and that available explanations are at the end of the day based on actually simple, elementary entities. True, explanations must in fact come to an end; and they stop at the level of powers whose mode of operation remains unknown, for the moment at least. Granted, such situation does not yield an argument in favour of the existence of powers. Yet the fact that science always seeks to discover explanatory mechanisms that involve categorical properties only in no way justifies declaring that powers do not exist.

5.7 For a Dualist Metaphysics of Laws of Nature

Most of the elements of the metaphysics of nature which I favour have been introduced earlier. It is time to bring them together while clarifying some important points. Like Armstrong,[41] I defend an ontology of states of affairs, that is to say facts,

[39] See Chap. 1, Section 1.6.

[40] This discovery was made in 2015 by the researchers of the *Laser Interferometer Gravitational-Wave Observatory* (LIGO). (Abbott, B. P. et al. 2016)

[41] This seems to be close to what Wittgenstein says in the *Tractatus* (1.1 "The world is the totality of facts, not of things" and 2 "(…) a fact is the existence of states of affairs"). (Wittgenstein 1971) Facts can be multiple states of affairs or single ones.

which are either thick particulars, denoted "F*a*", or relations, denoted "*a*R*b*".[42] Facts are instantiated properties or relations. Following Armstrong again, I am inclined to construe properties as universals, rather than tropes. Yet, unlike categorists, I include in my ontology both powers and categorical properties, instantiated in particulars. My ontology dispenses with irreducible natural kinds and the related distinction between essential and accidental properties.

The predicative terms in nomological statements refer to categorical properties. I submit that spacetime and structural properties do not exhaust the domain of categorical properties. All observable properties are categorical. Observable categorical properties make it possible to empirically individuate particulars whose states or behaviours satisfy scientific laws. Thus, the Boyle-Charles nomological formula is satisfied by "ideal" gases, under normal circumstances, and only by them. Gases are usually identified by ordinary categorical properties, such as their fluidity, tactile properties etc. The word "gas", which is a natural kind term, functions as an abbreviation to refer to a set of commonly co-instantiated properties. This does not imply that there is a *unique* set of properties necessary and sufficient to identify an individual as belonging to a specific natural kind. For example, electrons are the only entities that have a negative charge of 1.6×10^{-19} Coulomb. Thus, they can be identified by one property only, as well as their values of charge and mass, taken together. The natural kind terms denote entities that possess powers as well, which can also be called "potentialities" or "potencies". In the worlds where specific powers are instantiated they ground the necessary truth of laws and their associated counterfactuals.

Quantitative properties such as pressure, volume, charge etc. can take an indefinite number of values. These are determinable properties. As said above, Ellis claims that general quantitative properties have ontological priority over their particular values. Unlike Ellis and in agreement with Armstrong (and Aristotle...), I take particular values of mass for example to be ontologically more fundamental than the generic property of mass. It is true that the instantiation of a specific value of mass logically implies that the general property of mass is also instantiated, just as an instantiation of water implies the instantiation of oxygen and hydrogen, but not conversely. It is also true that oxygen and hydrogen are ontologically prior to water since a water molecule is made of one oxygen atom and two hydrogen atoms. But, unlike water, the generic property of mass is not composed or made of the totality of its instantiated values. It may indeed happen that some values of mass are never actually instantiated in our world and that there are so to speak "gaps" in the continuum of mass values. For the general property of mass to exist and be knowable, it is sufficient that some value of mass be instantiated, just as the reality of baldness (or to be bald, which is a universal property) does not require that all possible bald people exist at some time.

---

[42] Of course, besides dyadic relations with two argument places like *a*R*b* the ontology I propose also contains polyadic relations, with three, four etc. argument places.

Another significant issue is the nature of the connection between the nomicity of a universal proposition and the truth of counterfactual propositions. From what has been argued before, the basis for the necessary truth of laws of nature and their associated counterfactuals in some worlds are the instantiations of the appropriate powers in these worlds. If this is correct, both the nomicity of the proposition <all gases at equilibrium in normal (*ceteris paribus*) circumstances manifest the formula PV=KT > and the truth of the counterfactual proposition <if a gas were at equilibrium in normal circumstances, it would manifest the equality PV=KT > are founded on the instantiation of the appropriate natural power.

At this point, an important clarification is needed. As said, the possibility for a kind of particulars (sugar cubes, electrons, gases, charged entities and so on) to instantiate specific manifestations is warranted by the instantiation of some specific power. What about the necessity of a law? A law is distinct from a nomological formula since it explicitly or implicitly refers to the particulars and the circumstances in which these particulars will manifest some nomological formula. If the power identified by a specific nomological formula is instantiated in some particular, this particular will necessarily manifest the nomological formula if the conditions or provisos mentioned by the law are satisfied. This is what nomic necessity consists in. Such activation of power is not necessary. A power could be dormant. The occurrence of a manifestation often depends on property values which are referred to in the literature as *initial conditions* or *boundary conditions*. Those conditions obviously are contingent. In mathematized disciplines, such as physics, these appropriate conditions are often explicit. For example, the laws of mechanics remain approximately true in known circumstances as shown earlier.

When a mathematical formulation of a law is available, the initial conditions or boundary conditions are often mentioned in the nomological formula that describes the manifestation of a power. In many cases, the so-called triggering or activating circumstances of a power are referred to by symbols in those formulas. For example, velocity and position are referred to in the laws of motion and at the same time specific values of them are initial conditions. Coming back to the example of the sugar cube immersed in water, it is usually said that the triggering circumstance for the activation of the dissolution power is simply putting the sugar cube in water. But dissolution can happen in many ways: dissolution is a kind of processes. The precise description of a particular dissolution process by means of mathematical equations depends on the initial positions and velocities of the molecules of water and sugar. As a consequence, in many scientific cases, it is not possible to distinguish the triggering circumstances and the manifestation.

It is time to address the crucial issue of the reality of powers. How can we justify belief in their reality? For the dispositional realist, the facts that ground both the nomicity of laws and the truth of counterfactual propositions are (thick) particulars that instantiate powers: they are modal facts. But are we justified in claiming that such modal facts exist? What reasons can we adduce for believing in the existence of natural powers? Recall that we must distinguish between natural dispositions and pseudo-dispositions such as the capacity to drink water while running or to recite Apollinaire while repairing a flat tyre. The dispositions we are especially interested in are the powers that make entities capable of engaging in regular processes or persisting in states described by scientific laws, or more precisely, nomological formulas.

To prove the existence of powers, we must once again appeal to experience. At this point, we are immediately confronted to the impossibility of observing modal properties, which are, in this sense, metaphysical properties. Nevertheless, it is possible to verify that a sugar cube is soluble by immersing it in water. Admittedly, this is an *after-the-fact* verification that a particular instantiated a power when it no longer does. On the other hand, a proposition that attributes a power like solubility to an entity is falsifiable, and this immediately shows that this proposition is not empty or tautological. It may happen that an entity which looks like a sugar cube does not dissolve (under normal circumstances similar to those in which the dissolution of sugar has been observed) and thus does not possess the property of being soluble.[43] According to the correspondence view of truth (and falsity), there must be a reality – a falsifier – that makes the proposition <this is soluble> false. For the same reason there must be facts – truthmakers – that make such assertions true when they are. What are those truthmakers?

It is easy to ascertain that the categorical properties shared by sugar cubes (such as hardness, flavour etc.) have in the past been always associated with dissolution processes that have actually been observed. By induction we conclude that it is a law that sugar cubes immersed in water dissolve. (Recall that universal propositions are identified as scientific laws in the context of predictively correct theories.) We believe that sugar cubes will continue to behave in this way and thus that they have the possibility to dissolve. What is more, repeated experience instilled in us a sense of inevitability of the occurrence of the dissolution of sugar when some circumstances occur. David Hume would agree that such experience is the source of our idea of necessity. But since some universal propositions (laws) and their associated counterfactuals must have external truthmakers as argued earlier, I submit that our experience of inevitability counts as evidence for the existence of external entities, which are real natural powers.[44]

Then, it is because I know (in the sense of having good reasons to believe) from repeated experiences that sugar cubes are soluble that I keep them in dry places. It is because I know that the book I hold in my hand has the power to fall that I do not drop it. It is because I know that some vases are fragile that I handle them with care.

Once we have (indirect) empirical evidence for the instantiation of powers, i. e. dispositional facts that make some other facts possible, we also have empirical evidence in favour of the existence of facts which *conditionally necessitate* other facts. Since counterfactuals state that there is a necessary link between an antecedent and a consequent, their truthmakers must ground such necessity. Thus, I contend that we have indirect evidence in favour of the existence of powers which provide the

[43] The objections based on the possibility of finks and antidotes have been examined earlier (Sect. 5.3). Normal circumstances exclude the presence of finks and antidotes.

[44] See Ghins (2007, pp. 143–144).

ground for the necessary connection between the antecedent and the consequent in counterfactuals when they are true.

Notice, and this is an important point, that we do not infer the existence of powers by invoking an inference to the best explanation of the observations.[45] Certainly, instantiations of powers make it possible to give an explanation of certain regularities, an explanation which admittedly is not scientific but metaphysical. But the argument proposed above in favour of the existence of powers is a *bottom-up,* not *a top-down*, argument. It is because we have repeatedly seen that sugar cubes dissolve in some circumstances that we conclude that it is a law that sugar cubes always and necessarily behave in such a way. Hence, we are entitled to claim that a specific counterfactual <if I had immersed this sugar cube in water, it would have dissolved> has a truthmaker. We also noticed that counterfactuals can be falsified by observations. We are then in a position to defend that some dispositions are not just "inference tickets"[46] but that real powers do exist, not because they provide an account of the truth of some counterfactuals (of course, they also do), but on the basis of observations.

Within the framework of such inductive epistemic approach, the starting point is provided by observations: we have a bottom-up argument. Conversely, from an ontological point of view, powers are prior to observations: they are the ontological foundation of the conditional necessity of observed regularities described by laws and, at the same time, of the truth of counterfactual propositions.[47]

It seems to me however that the most convincing experiences in favour of the existence of powers are my own personal experiences. I know that I can get up from my seat now if I choose to do so. Yet is it legitimate, on the basis of first-person experiences, to attribute powers to external entities by some sort of anthropomorphic transfer? Projecting our private experience onto external things by attributing to them a dynamic principle is a highly objectionable move for sure. Nonetheless, if external things that resemble each other react to my actions in a regular specific way, by inductive reasoning it seems legitimate to predict that they will continue to do so and thus that they have the real possibility to behave thus. This amounts to attributing dispositions or powers to them.

Once the existence of powers has been attested,[48] we get a solution of the ontological problem of identification of laws. Indeed, a law of nature is a universal

---

[45] Contrary to what Vetter (2015, p. 12) and others suggest.

[46] As Gilbert Ryle famously contended.

[47] As Stavros Ioannidis pointed to me, Humeans (such as Gilbert Ryle) could accept that sugar cubes do have the disposition to dissolve without having to postulate the existence of powers in a substantial sense. My argument however decisively relies on the truthmaker principle. There must be something real, namely powers, to ground the truth of some counterfactuals without having to resort to the heavy metaphysical supposition of concrete possible worlds, as Lewis does.

[48] Notice that I do not resort to elementary particle physics to vindicate the existence of irreducible powers. This naturalistic argument is defended by, among others, Harré and Madden (1975, pp. 154–155) and also Molnar (2003, pp. 135–137). For a critique of this argument, see Psillos (2006, pp. 172–174). As I have already said, I am not a supporter of naturalized philosophy,

proposition that states that under normal conditions (*ceteris paribus*), some particulars – such as gases, electrons etc. – manifest behaviours or states satisfying specific nomological formulas. The nomicity of such propositions is based on the instantiations of appropriate powers. Thus, natural or metaphysical necessity, which is grounded on the existence of powers, does not coincide with logical or analytical necessity. It is a *de re*, not *de dicto,* necessity.

As Mumford puts it:

> The metaphysically possible words are plausibly, therefore, a sub-set of the logically possible worlds, simply because the constraints on logical possibility are so weak. (...) It is wholly plausible, therefore, to speak of something being metaphysically necessary but logically contingent. (. . .) we cannot deny that there are different *modes* of necessity. (. . .) A notion of contingent necessitation is not so hopeless, as long as we are not saying that something can be contingent and necessary in respect of the same mode of necessity. There was no reason to think Armstrong and other contingency theorists were ever claiming something quite so contradictory. (Mumford 2004, p. 165)

Let us take stock at this point. Unlike Armstrong, I have argued that laws are not realities but propositions. In agreement with Mumford, I contend that the necessity of nomological propositions is not logical nor analytical. Since the instantiation of a power in a particular necessarily implies the occurrence of regularities *if* certain conditions are fulfilled, this is sufficient to ground the necessity of laws and the truth of counterfactual propositions in our world, without having to resort to the concrete existence of possible worlds and relations of proximity between them, as Lewis proposed. The instantiation of a power in a particular readily guarantees that it would behave in accordance with a nomological formula in case the favourable circumstances are realised in our world (which is the only real one after all. . .) Instantiated powers are thus partial truthmakers of counterfactuals and the basis of their necessity. At the same time, powers ground the truth of statements of the form "p is a law".

A standard objection to this view is that the nomicity of *conservation laws* cannot be grounded on powers alone. (Bird 2007; Livanios 2010, 2017; Ioannidis et al. 2021) A conservation law states that specific quantities such as energy, mass, charge etc. remain invariant for *closed* systems. Generally, conservation laws follow from the invariance of the mathematical form of dynamical laws under some group of transformations (also called "symmetries" of the laws).[49] For example, the conservation of energy is a consequence of the invariance of the mathematical forms of the laws under time translations. Yet conservation laws do not only hold for individual

---

whatever form it may take. Science, including physics, is – and always will be – unable to support arguments for or against genuinely philosophical, and therefore normative, positions.

[49]For continuous transformations we also speak of Lagrangian symmetries. A Lagrangian function for a physical system is usually the sum of its kinetic energy and an interaction term (potential energy). The Euler-Lagrange equations are differential equations which are obtained from the Lagrangian and the Hamilton's principle of extremization of the action. From these equations, we can derive by integration the laws of motion of masses in a gravitational field, for example. For a detailed and illuminating discussion of Noether's theorems and the connection between Lagrangian symmetries and conservations laws see Brown (2022).

particulars (such as individual particles) but also for *systems* composed of several particulars, whereas powers are instantiated in individual particulars. What is more, conservation laws are satisfied by *closed* or *isolated* systems only. Since the unique strictly closed system is the whole world, conservation laws cannot possibly be grounded on *local* powers.[50] Thus, conservation laws do not describe possible manifestations in particulars endowed with some powers, which are supposed to be the sole ground of the necessity of those laws.

My reply to this objection goes like this. First of all, individual components of a system all instantiate some powers from which the behaviour of the whole system depends. Take the example of two positively charged bodies interacting in accordance with Coulomb's law. Each charge has the power to repel the other one. The manifestation of the repulsive power is the electrostatic force. Admittedly, such manifestation is relational since it involves two charges, but this does not prevent the repelling power from being intrinsic to the individual charges. If the system is isolated and no other forces are present, the two bodies will move away from each other according to Newton's second law. As we saw, Coulomb's and Newton's laws (in fact, nomological formulas) describe the manifestations of powers instantiated in the particular bodies. The closure of the system – which of course is never perfect – is part of the normal conditions on which the manifestation of a power depends as well as the (approximate) satisfaction of the law. (Approximate) isolation is a local condition that does not require the consideration of larger systems or the whole universe as some have claimed.

Since the manifestations of powers are described by nomological formulas, their mathematical symmetries are part and parcel of the descriptions of manifestations. In the example above, invariance under time translation implies the conservation of the energy of the system (if it is – approximately – closed). Moreover, the (approximate) empirical correction of nomological formulas for Coulomb's electric force and Newton's second law imply that the involved quantities of charge, and mass are conserved.

Just as we inductively generalise the validity of some laws and hold them to be true for the entire world, I see no reason to block the same move when it concerns the formal symmetries of laws. If this is correct, global symmetries are empirically justified by inductions starting from local observations. Notice further that such reasoning is perfectly in harmony with an empiricist epistemology.

Conservation laws and symmetry principles, which could be called "meta-laws",[51] thus fully deserve the title "law of nature". Their truth is grounded on the

[50] This is why Bigelow, Ellis and Lierse (1994) have sustained that conservation laws flow from the kind-essence of the whole world. However, their proposal is not immune of the difficulties that beset the notion of essence and the problematic empirical meaning of the "kind-essence of the world". Moreover, they argue within an explanationist framework, whereas I am an inductivist as I spelled out before. At any rate, their explanation of the origin of the necessity of conservation laws is not satisfactory, as Ioannidis et al. (2021) rightly claim.

[51] This expression is also used by Lange (2000) but with a different meaning.

instantiations of intrinsic powers. If I am right on this, powers (not laws) do govern the behaviour of things.[52]

It remains to examine the question of the nomicity of the so-called "empty" laws, i.e. laws that are not satisfied by regularities actually occurring in nature. A typical example is that of the classical law of inertia. Since, in fact, a body is always subject to forces whose sum is different from zero, we will never observe a perfectly rectilinear and uniform motion.[53] Yet we observe many regularities that come very close to the ideal processes described by such laws. Moreover, we know that the observed deviations from what is stipulated by a law are due to well-identified factors. From experience, we have good reason to suppose that if these factors had been absent we would observe regularities exactly conforming to those described by the nomological formulas.

David Armstrong (1983, pp. 147–150) distinguishes between *iron laws* and *oaken laws*. Iron laws do not mention any *ceteris paribus* conditions and are therefore exactly true and verified (leaving aside possible errors of observation). For example, the law that describes the rate of decay of radium does not depend on external circumstances: it is an iron law. Another example is Coulomb's law. If two charges are present, they necessarily attract or repel each other. Some circumstances may prevent those charges to move, but not their attraction or repulsion in concordance with Coulomb's formula.

Oaken laws are those that mention explicitly or presuppose *ceteris paribus* conditions. If a well-confirmed law appears to be falsified, this is most often due to the presence of disturbing factors. In some cases, however, falsification will lead to restricting the domain of validity of the law according to factors unknown until then, which would amount to reinforcing the *ceteris paribus* conditions by making them more restrictive. I have already mentioned the case of the laws of classical mechanics which remain true (to a very good approximation) if velocities are small compared to that of light, if the gravitational field is weak and if the action is large with respect to the value of Planck's constant. The possibility that new restrictions may be discovered later is left open. Such an approach does not amount to protect the laws of classical mechanics against falsification whatever happens but to delimiting their truth domain, which can also open the door to the discovery of new laws that would hold true for novel empirical domains. The abundant literature of the last quarter of the twentieth century (Lakatos 1978; Leplin 1997) made it clear that the *ad hoc* manoeuvres concocted to try to "save" the truth of a law in all circumstances are not admissible. Accurate empirical methods make it possible to determine the

---

[52]This is why I cannot follow Ioannidis et al. (2021) who contend that powers alone are unable to ground the necessity of conservation laws and symmetry principles and more generally to perform a governing role.

[53]In his general theory of relativity, Einstein showed that the motion of a body subject to gravitational forces alone could be considered inertial since, in this theory, inertia is incorporated into the structure of spacetime. But such inertial motion is not rectilinear in the Euclidean sense. However, the classical law of inertia remains true to a very good approximation in many contexts.

conditions under which a law remains (approximately) true and to explicitly include them in the *ceteris paribus* conditions.

With this, I conclude the presentation of the metaphysics of laws of nature grounded on causal powers and categorical properties I propose. In order to forestall several objections, I must now come back to two important issues, namely the relations between categorical properties and powers and the kind of necessity enjoyed by laws of nature. This is the aim of the following two sections.

5.8 What Is the Relation Between Powers and Categorical Properties?

Dispositional monists, such as Stephen Mumford and Alexander Bird, will object to proponents of a dualist metaphysics of both powers and categorical properties that the so-called "categorical" properties are in fact hidden dispositions. In other words, there are no properties that are genuinely categorical. It is true that at first sight properties like charge and mass appear to be powers. If a body is charged, then it has the possibility of behaving in accordance with Coulomb's law in the presence of an electric field. Charge however is an observable property in the broad sense, and therefore categorical. As explained before, straightforward experience (for example, with an electroscope) permits us to see that charges are present. True, we know today that the observation that a body is charged is made possible by Coulomb's law. But this is an epistemic point. Such observation is independent of the existence of the power of attracting and repelling other charges. Therefore, it is essential to distinguish the categorical property of charge from the *power* of attracting and repelling and furthermore of behaving in accordance with Coulomb's law. The predicative expressions "to be charged" or "to have in charge of a specific value" on the one hand, and "to have the possibility of exerting a force in conformity with Coulomb's functional quantitative law" on the other hand, have distinct referents: the first denotes a categorical property, the second a power. Here again, language is playing tricks with us if we use the word "charge" to indistinctively refer a categorical property and a power.

This same remark applies to the property of mass. Interestingly, in his *Principia,* Newton (1687) first defines (*Definition I*) the "quantity of matter" as the product of density and volume. Such definition is clearly categorical. Then, in *Definition III*, he defines the "*vis insita*" or "innate force of matter" as a "power of resisting", which corresponds to our inertial mass.[54] Of course, we commonly use the same word "mass" to refer both to the quantity of matter and the inertial mass. Yet the first is a categorical property, whereas the latter is a power.

Granted, there is a close connection between powers and categorical properties. First, the former ones are individuated by means of the latter. The manifestation of a

---

[54] I am grateful to Theodoros Arabatzis for having pointed out this to me (private communication).

power as well as the conditions of its manifestation are described by categorical properties. What is more, a particular which instantiates a power will also instantiate categorical properties, some of which are necessary for the instantiation of that power. The capacity to be attracted by a positive charge for example necessarily requires the instantiation of the categorical property of being negatively charged. Such necessity is contingent. In possible worlds where charges of the same sign attract each other such manifestations correspond to powers distinct from the ones instantiated in our world.

If we include categorical properties in our ontology, it is possible to refute a classical objection against the existence of irreducible dispositional properties. David Armstrong formulates it as follows:

> Given a purely Dispositionalist account of properties, particulars would seem to be always re-packing their bags as they change their properties, yet never taking a journey from potency to act. For "act", on this view, is no more than a different potency. (Armstrong 1997, p. 80)

As Stathis Psillos (2006, p. 162) puts it, "If properties are nothing but powers, then when a power is manifested, its *effect* (the acquiring of a property by a particular) will also be a *power*. Hence, nothing really happens apart from the shifting around of powers from particular to particular."

Molnar (2003, p. 173) dubbed this argument "*always packing, never travelling*". Here, dispositional monists, like Molnar and Mumford, are confronted with a difficulty.[55] Indeed, powers allow nature to be set in motion. Yet if the only possible movement is that of a new distribution of powers among particulars, the world as a whole would appear to be completely motionless since powers are not observable. Nevertheless, powers are actual and real even when the particulars in which there are instantiated do not manifest them. Thus, Armstrong's argument is far from dealing a fatal blow to the existence of powers, since, according to power monists, a particular does change when it acquires a power it did not instantiate previously. When a change occurs, a particular that had a power P acquires a power P′. For the dispositional monist something really happens. Actual movement in nature takes place, whether we are aware of it or not. But in a world in which only powers exist, there are no instantiations of *observed* properties and thus no observed change. Since powers are neither immediately observable nor detectable, we should not observe any change. Such conclusion is of course unacceptable, and the dispositional monists find themselves in a serious predicament. A way to get out of such predicament would be to claim that powers are observable after all. But such a move is highly counterintuitive and therefore unacceptable.

From an epistemic point of view, the presence of a specific power is attested, indirectly by relying on its manifestations and the truthmaker principle, by the intrinsic similarity between, for example, sugar cubes whose dissolution has been

[55] See Mumford's response (2004, pp. 174–175). He does not address the epistemological predicament, however.

observed and those that have not been immersed in a liquid. Such inductive empirical evidence consists, once again, in the observation of categorical properties.

In the *Sophist*, Plato puts the following sentence into the mouth of the stranger from Elea:

> I suggest that anything has real being that is so constituted as to possess any sort of power either to affect anything else or to be affected, in however small a degree, by the most insignificant agent, though it be only once. I am proposing as a mark to distinguish real things that they are nothing but power. (247e; Plato 1973, p. 992)

As said earlier, I claim that every being, every particular, has intrinsic powers and that, in order to be, a power must be instantiated. But I subscribe only partially to what has been called the "Eleatic Principle", as formulated by Plato above, since according to me real things (particulars) are not just (instantiated) powers. A particular also has categorical properties, the existence of which is testified in actual observation. Powers do not exhaust the domain of real properties, contrary to what dispositional monists maintain.

5.9 What Kind of Necessity for Laws?

Necessity is the most important distinctive feature of laws. As I have argued, the necessity of nomological propositions is founded on the instantiation of powers. Thus, the truth of a proposition <p is a law of nature> where "p" denotes a scientific law,[56] is grounded on real powers instantiated in the particulars to which scientific laws apply (gases, charged bodies etc.). Accordingly, such a necessity is conditional upon the contingent instantiation of powers in a subset of logically possible worlds.

The necessity of scientific identifications such as < water is H_2O > or < a gas is a collection of molecules in motion> is not analytical. These are not definitions but *de re* necessary propositions. As I stressed, a gas is first defined as an entity with certain superficial, observed properties. Since some laws link these surface properties to microscopic properties we can *redefine* gases on the basis of the latter. Such redefinition does not add any new content to the laws that state necessary causal connections between the surface properties (the effect) of a gas and its microscopic properties (the cause). The same holds true for the scientific redefinitions of water, the electron[57] and so on. Since these identifications are consequences of laws of nature, they are necessary as well.

[56] Remember that a scientific law is a universal proposition belonging to a well confirmed scientific theory and which we have good reasons to believe true.

[57] Remember that the electron is first identified as the cause of optical phenomena in discharge tubes (see Chap. 3). Necessary laws link optical phenomena to intrinsic properties of electrons.

It might be that in other logically possible worlds, particulars with surface properties identical to those of our gases would be in a state of equilibrium that satisfies the equality

$$PV = K^*T$$

where K* differs from the value of K in our world (and the categorical properties of pression, volume and temperature are the same as in our world). According to the original description of the term "gas" the proposition <all gases at equilibrium satisfy the equality PV=KT > is false and cannot be a law in such a world. Essentialists, such as Brian Ellis, would reply that gases form a natural kind and that in all possible worlds in which there are gases, they must conform the Boyle-Charles formula, when in a state of equilibrium. If not, we are not dealing with the natural kind gas but with some other kind. In a world where the kind gas is not instantiated, the law of ideal gases is true, but in an "empty" way (Ellis 2002, p. 101).

This position is defensible for essentialists such as Ellis for whom a law is a description of the essential properties of a natural kind. It is true that such description remains correct even if the natural kind is not instantiated by any particular. However, this contention depends on the acceptance of substantial natural kinds, the existence of which does not depend on the existence of its associated class, as pointed out before. For those who do not admit the existence of substantial natural kinds, like Armstrong and me, this move is not permissible. The proposition <all gases at equilibrium satisfy the equality PV=KT > is false in a world where gases, according to the initial description, exist and behave in conformity with the equality PV=K*T.

In our world, the fact that a particular instantiates a power in some location and at a certain time is obviously a contingent fact. That an object, like a book, is now on my desk is a contingent fact, regardless of whether there are concrete possible worlds *à la* Lewis in which this book is in another space-time location or in which there are no massive bodies. From the occurrence of this fact, it follows – necessarily – that certain other facts will take place, if certain conditions obtain. If I push it towards the edge of the desk it will fall. The book has the possibility to fall, even if I do not push it. *De re* possibility is a notion which is easily grasped from experience in the first place and not from logic. No need to resort to possible worlds. To understand possibility, and modality in general, we can dispose of what Peter Verdée rightly calls the "monstrous concept"[58] of an infinite number of concrete possible worlds the existence of which is evidently not supported by any observation. The understanding of the notion of *de re* possibility is based on the observations of what I can and cannot do and what can and cannot happen in our world. Logical and analytical *de dicto* necessity, which is purely formal or semantic, is impotent to ground real modalities. There is no logical contradiction in my levitating now. But I clearly know that that there is no possibility for me to do so.

---

[58] Personal communication. I am grateful to Peter Verdée for helpful comments on this section.

5.10 Conclusion

Summarising, I tried to defend a dualist metaphysics of nature that includes both categorical and irreducible dispositional properties, namely natural causal powers. If particulars instantiate real powers, this solves the ontological problem of identifying the grounds of the *de re* necessary truth of laws of nature. Such mode of necessity is contingent upon the instantiation of powers. Moreover, the empiricist requirement of observability led me to claim that particulars also instantiate categorical properties. Unlike most authors, I argued that observable properties, which are categorical properties, are not quiddities.

Empiricist philosophers such as Bas van Fraassen (1989) have raised sharp objections against an ontology of powers. For them, the absence of direct empirical access to modalities in general, and *a fortiori* powers, undermines the foundations of such an ontology. What is more, van Fraassen (2002, p. 10) rightly argues that it is illegitimate to solve philosophical problems by postulating the existence of empirically inaccessible entities.

The recourse to powers solves the ontological problem of identification of laws. Nonetheless, this explanationist argument in favour of the reality of powers is unconvincing. This is why I attempted to show that an indirect recourse to experience legitimates belief in their reality. According to the dualist ontology of properties I propose, the world consists of facts that are instantiations of properties and relations. There are two types of facts: instantiations of categorical properties and instantiations of powers. In this sense, we have an ontology of facts, either categorical or modal. Such ontology in no way prevents particulars from co-instantiating the two kinds of properties.

The laws of nature are construed as necessary propositions, in the mode of *de re* natural necessity, which is distinct from and more restricted than logical necessity. Powers, not laws, are the driving force for change in nature and govern the regularity of processes described by nomological propositions. Powers endow the particulars that instantiate them with some dynamism, i.e. the possibility to engage in a specific behaviour. If certain contingent circumstances occur, a particular necessarily instantiates or manifest the behaviour corresponding to the power it is endowed with. This is why the world is not composed of a mosaic of separate and mutually independent facts, contrary to what Wittgenstein contended in the *Tractatus*. Powers are the grounds of the necessary connections between the properties which are their manifestations, which are described by nomological formulas. The laws of nature are propositions that describe the behaviours that specific particulars will necessarily manifest when the appropriate circumstances obtain.

Chapter 6
General Conclusion

To conclude, I will first recall the central theses of the metaphysics of science that I defend. I will then mention some difficulties of my views, which hopefully could provide incentives for further research.

If there are no cogent reasons to believe in the (at least partial and approximate) truth of our best scientific theories, it would be futile to develop an ontology and *a fortiori* a metaphysics of science. This is why I have argued at length for a selective epistemological scientific realism according to which we have better reasons to believe in the reality of certain (directly or indirectly) observable properties and relations instantiated in particulars, than to suspend our judgement with respect to their existence. These properties are organised into structures represented by the models of our theories. In order to remain faithful to a moderate empiricist epistemology, I advocated a version of scientific realism based on a bottom-up inductivist strategy, distinct from the top-down explanationist strategy usually mobilised by scientific realists.

Scientific laws are propositions that describe regularities, i.e. temporal successions of facts, which are understood as instantiations of properties. The first problem we face is to define a criterion in order to determine if a universal proposition is a scientific law rather than a proposition which describes a fortuitously occurring regularity. This is the *epistemological* problem of identification of scientific laws. The second problem hinges on what are the real grounds of the necessary truth of a scientific law such as <under normal circumstances, all (perfect) gases at equilibrium manifest a state satisfying the Boyle-Charles formula>? This is the *ontological* problem of identification.

I proposed to bestow the title of *scientific* law only on universal propositions belonging to a predictively correct scientific theory and which we have excellent reasons to believe to be true, with the exclusion of tautologies and analytic propositions. For this criterion of scientific nomicity to be applicable, we must first agree on what a predictively correct scientific theory is. Since the so-called "problem of demarcation" (between scientific and pseudo-scientific theories) and the problem of confirmation of theories have been the subject of an abundant literature throughout

M. Ghins, *Scientific Realism and Laws of Nature: A Metaphysics of Causal Powers*, Synthese Library 483, https://doi.org/10.1007/978-3-031-54227-5_6

the twentieth century, I did not deem it opportune to return to them. I simply considered as scientific the theories that are almost unanimously recognised as such. Nonetheless, I offered a synthetic conception of theories according to which scientific laws are propositions, in accordance to the syntactic view. However, laws are made true by models that represent structures of properties, in line with the model-theoretic view of theories.

Once the scientific laws have been epistemically identified, we must determine the kind of external realities which ground their nomological status, that is, their necessity, and to resolve the ontological problem of identification. Following Stephen Mumford, Brian Ellis, Alexander Bird, Anjan Chakravartty, among others, I defended the existence of irreducible dispositional properties, that is, more precisely, natural causal powers. If a particular instantiates a power, it will necessarily manifest a specific behaviour – described by what I have called a "nomological formula" – provided that the appropriate "normal" circumstances are present.

What ontologically distinguishes nomological regularities from fortuitously occurring regularities is that the former are manifestations of natural powers. A law of nature is a scientific law whose nomicity is based on the instantiation of natural powers. Remember that a condition of adequacy for a philosophical theory of laws is that laws must be capable to support the truth of counterfactual conditionals. In my view, if counterfactuals can be supported by laws it is because their necessary truth is grounded on the same metaphysical properties that make laws necessary, namely causal powers. Particulars which instantiate intrinsic powers are modal facts which are partial truthmakers of laws of nature and also of the counterfactual propositions corresponding to them. The counterfactual proposition <if this metal were heated, it would expand> expresses a necessary link between the antecedent and the consequent. Such *de re* necessity – as opposed to analytical *de dicto* necessity – is based on the existence of a specific power instantiated in metals. Since the spatiotemporal instantiation of a power is a contingent fact, nomic natural necessity is relative to the worlds in which such instantiation occurs. Nomic necessity is more restricted than logical necessity. The laws of nature, which are necessarily true in our world, may not be necessarily true in other logically possible worlds. Accepting that metals, identified by means of the same observable properties as in our world, could behave differently in other possible worlds, does not involve any contradiction. On this, I disagree with dispositional essentialists such as Brian Ellis and Alexander Bird who claim that laws of nature are metaphysically necessary in an absolute (which they call "metaphysical") sense, that is, not relatively to some worlds.

Ontology has to do with the realities that populate nature. As I said, I support an ontology of instantiated properties, i.e. of facts. Since powers are modal properties, they are not observable: we observe the dissolution of sugar in a liquid, but not its possibility of dissolving. Being unobservable, powers are metaphysical properties. However, I have tried to make a case in favour of their existence by relying on the truthmaker principle and implementing an inductive empirical strategy. In my understanding of the term "metaphysics", an ontology that only includes observable entities is not a metaphysics. On the other hand, an ontology that includes powers, potentialities or potencies (whatever one calls them) is a metaphysics of nature.

If a property is observable it is categorical. Categorical properties and dispositional properties form distinct classes: a categorical property cannot be a dispositional property, and vice versa. Empiricist philosophers inspired by David Hume have attempted to articulate conceptions of laws that rely exclusively to categorical properties. These Humean philosophers are either regularists like Mill, Ramsey and Lewis, or categorical necessitarians like Armstrong, Dretske and Tooley. However, I have followed Bas van Fraassen and also Stephen Mumford and Alexander Bird, among others, and argued that these conceptions are unable to simultaneously solve the two problems that any philosophical theory of laws must be able to resolve: the inference problem and the identification problem. While regularists easily manage to provide a straightforward solution of the problem of inference, they fail to identify the kind of realities in the world which ground the necessity of nomological propositions. The categorical necessitarians, on the other hand, adopt an ontological conception of laws and identify them with a real relation of necessitation between categorical properties. For them however, the inference problem as well as the ontological identification problem remain major hurdles that they have been unable to overcome so far.

For my part, I have defended a metaphysics that includes natural powers, which are similar to Aristotle's potencies. Alongside powers, and treading in Brian Ellis' footsteps, I also include categorical properties in my ontology. I thus advocate a dualist metaphysics of nature in opposition to dispositional monists, such as Stephen Mumford and Alexander Bird, who believe that the so-called categorical properties are disguised dispositions. In my view, the recourse to categorical properties is crucial if one aims at developing an ontology of laws based on observations. Classically, only spatial, geometrical and structural properties are considered to be categorical properties. Yet I contended that properties like mass and charge can also be deemed observable, at least in principle. Unlike Ellis and others, I then recommended to extend the domain of categorical properties to all observable properties and not to limit them to geometrical and structural properties only, as empiricists usually do. What is more, I do not think that categorical properties are quiddities enjoying just a primitive identity. The same applies to all the observable properties mentioned in laws. I have argued that the instantiation of a disposition in a particular necessarily involves the instantiation of categorical properties, without which the disposition could not be present. A categorical property can play a function, a causal role, which is always the same whatever the world in which it might be instantiated, without itself being a causal power. However, I acknowledge that the articulation between categorical properties and powers remains a vexing problem, to which I do not claim to have provided a satisfactory solution.

A power endows the particular in which it is instantiated with real capacities of action, that is the real possibility of manifesting causal processes or persisting in states described by nomological formulas. Thus, planets manifest specific regular motions, namely temporal successions of positions and velocities, described by Kepler's formulas. The ideal gas law, which is a state law, describes the manifestation of the power to persist in a specific state of pressure, volume and temperature when isolated and at equilibrium. The world is made of categorical, inert and passive facts, but also of dispositional, active and dynamic facts. In a regularity, facts follow one another by virtue of the instantiation of internal powers. In contrast to the Humean view, for which nature is a mosaic of facts without real links between them, the neo-Aristotelian conception sees in the succession of regular facts the manifestation of powers that cement relations between these facts.

However, the metaphysics of laws I propose does not include important Aristotelian ingredients, namely essences and the *sui generis* existence of natural kinds. On this I disagree with philosophers such as Mumford, Ellis, Lowe, Bird and Chakravartty who maintain that natural kinds are irreducible realities. I do not think it to be possible to draw a satisfactory distinction between essential and accidental properties. I also believe in the possibility of reducing natural kinds to conjunctions of properties as Armstrong suggests, while recognising that such reduction is not free from difficulties.

The metaphysics of nature that I propose should be developed and clarified on some points, such as the articulation between powers and categorical properties, as we have just seen. But a major problem that deserves to be further investigated has to do with the ontological identification problem. As we saw, the identification of laws in the context of scientific theories solves the epistemological identification problem. Although it lacks direct empirical meaning, we have a sense that some natural regularities inevitably occur and that it would be vain to try to modify or counter them. Such sense of inevitability leads us to attribute some kind of necessity to the laws that describe those regularities. Despite of this, strict regularists such as Quine refuse to attribute any necessity to laws. But empiricists like David Hume and David Lewis grant that laws of nature are necessary. For Hume, such necessity resides in our minds while Lewis identifies laws as universal propositions belonging to axiomatized empirically successful theories which achieve the best balance between simplicity and strength in our world. Lewis' treatment of counterfactuals then depends on the actual existence of logically possible worlds and a problematic relation of proximity between our actual world and some possible worlds having the same laws as our world. Powers are then accounted for by means of counterfactuals. Yet it is unclear how possibility and necessity can spring out from Lewis' treatment of counterfactual conditionals.

I stressed that I reject the reification of possible worlds, while acknowledging their importance for semantics and logic. Nevertheless, no satisfactory ontology of laws can be developed without resorting to a metaphysics of modalities, especially the modality of possibility by means of which the other modalities can be defined. Barbara Vetter proposes to ground all modalities on real dispositions, which she calls "potencies", heralded as primitive entities identified by their manifestations only, independently of the context and circumstances which could activate their manifestations. I am sympathetic to this view which is quite in agreement with the neo-Aristotelian metaphysics of science I have presented. It works well for iron laws the necessity of which is grounded on dispositions which are instantiated to the "maximal degree" (Vetter 2015, p. 86). A disposition is maximal if it is always manifested by the particular which instantiates it, whatever the circumstances. Charges and samples of radium instantiate such maximal potentialities. But what about oaken laws? It seems somewhat radical to confer the title of law of nature only to the laws whose necessity is grounded on maximal potentialities.

We know from experience that the exercise of most powers depends on circumstances described by categorical observable properties. We also know from experience that if the appropriate circumstances are in place, we have a sense of ineluctability of the occurrence of some manifestation, the latter being again described by categorical properties. However, the existence of powers construed as endowing the mere possibility of a manifestation is not sufficient to provide the sought after truthmaker of assertions such as <this is a law>. The complete real grounding of the necessity of laws has not been provided because the ontological basis of the necessary link between the appropriate circumstances and the manifestation of a power has not been elucidated. Some clarification of the *de re* necessary connection between the appropriate circumstances and the manifestation of a power is needed. In my opinion, further research in this direction promises to be fruitful.

Another tricky issue concerns the nature of properties: are they universals or tropes? Fortunately, we can abstain from taking sides on this point, as Anjan Chakravartty suggests, since what is crucial for the metaphysics of nature I support is the reality of powers, regardless of whether these are construed as universals or tropes. This is true, but the metaphysician is then unable to specify the nature of powers, and thus of the realities that found the necessity of laws. I am leaning towards universals, but I have no strong argument to offer in their favour.

Also, the issue of laws in biology deserves to be thoroughly investigated. Some philosophers, like Stathis Psillos, maintain that there are no biological laws. This view appears to be exaggerated. It seems reasonable to uphold that disciplines like biochemistry and genetics contain laws. On the other hand, when we look at the theory of evolution of species, things become more complicated, mainly because the central concept of species is notoriously difficult to clarify. Defenders of biological laws, such as Chakravartty, rely on the notion of *cluster kind* whose members sufficiently resemble each other without it being possible to determine a set of essential properties common to all members of the same species and possessed only by them. I am myself sceptical about the possibility of identifying universal relationships between instantiated properties in the field of the evolution of species. The same is true, with a few exceptions,[1] for theories in social sciences. Be it as it may, the existence of laws in some disciplines remains an open question.

The metaphysics of science remains, even today, strongly influenced by a dualist conception of nature which goes back to René Descartes. For him, extended matter is inert and natural entities are passive. Natural motion is caused by divine external action, which requires material bodies to obey laws of conservation, in compliance with the immutability of God. Both regularists and categorical necessitarians endorse the Cartesian thesis of the passivity of nature, which has been taken over by Hume, according to which natural entities are intrinsically inert. This view of nature is challenged by a metaphysics of powers for which all changes and motions have their

[1] See Blondeau and Ghins (2012).

source in natural entities that instantiate powers, i.e. internal principles of action. Although it is unjustified to attribute to entities, as they are objectively studied in science, a kind of intentionality akin to mental intentionality,[2] a metaphysics of powers opens a path of research, which deserves to be explored, towards an overcoming of Cartesian dualism, which clearly separated the world of passive extended entities (substances) from that of thinking, intrinsically active, entities.

[2]On this question see Bird (2007, pp. 114–131).

References

Abbott, B.P., et al. 2016. Observation of Gravitational Waves from a Binary Black Hole Merger. *Physical Review Letters* 116 (061102): 1–16.
Achinstein, P. 2001. *The Book of Evidence*. Oxford: Oxford University Press.
Agazzi, E. 2014. *Scientific Objectivity and its Contexts*. Cham: Springer.
———., ed. 2017. *Varieties of Scientific Realism: Objectivity and Truth in Science*. Cham: Springer.
Alai, M. 2012. Levin and Ghins on the No-miracle Argument and Naturalism. *European Journal for Philosophy of Science* 2: 85–110.
———. 2014a. Novel Predictions and the No-miracle Argument. *Erkenntnis* 79: 297–326.
———. 2014b. Defending Deployment Realism against Alleged Counterexamples. In *Defending Realism. Ontological and Epistemological Investigations*, ed. G. Bonino et al., 265–290. Berlin: De Gruyter.
———. 2017. The Debates on Scientific Realism Today: Knowledge and Objectivity in Science. In *Varieties of Scientific Realism*, ed. E. Agazzi, 19–47. Cham: Springer.
Arabatzis, T. 2006. *Representing Electrons. A Biographical Approach to Theoretical Entities*. Chicago: University of Chicago Press.
———. 2012. Experimentation and the Meaning of Scientific Concepts. In *Scientific Concepts and Investigative Practice*, ed. F. Feest and F. Steinle, 149–166. Berlin: De Gruyter.
Armstrong, D. 1978. *Universals and Scientific Realism. Vol. I: Nominalism and Realism. Vol. II: A Theory of Universals*. Cambridge: Cambridge University Press.
———. 1983. *What is a Law of Nature?* Cambridge: Cambridge University Press.
———. 1989. *Universals. An Opinionated Introduction*. Boulder: Westview Press.
———. 1993. The Identification Problem and the Inference Problem. *Philosophy and Phenomenological Research* 53: 421–422.
———. 1997. *A World of States of Affairs*. Cambridge: Cambridge University Press.
———. 1999. The Causal Theory of Properties: Properties According to Shoemaker, Ellis, and Others. *Philosophical Topics* 26: 25–37.
———. 2000. Difficult Cases in the Theory of Truthmaking. *The Monist* 83: 150–160.
———. 2004. *Truth and Truthmakers*. Cambridge: Cambridge University Press.
Ascher, M. 1995. Models and Maps from the Marshall Islands: A Case in Ethnomathematics. *Historia Mathematica* 22: 347–370.
Bartels, A. 2005. *Strukturale Repräsentation*. Paderborn: Mentis.
———. 2006. Defending the Structural Concept of Representation. *Theoria* 55: 7–19.

M. Ghins, *Scientific Realism and Laws of Nature: A Metaphysics of Causal Powers*, Synthese Library 483, https://doi.org/10.1007/978-3-031-54227-5

Baum, R., and W. Sheehan. 1997. *In Search of Planet Vulcan. The Ghost in Newton's Universe.* Cambridge, Ma.: Basic Books.

Beebee, H., and J. Dodd, eds. 2005. *Truthmakers. The Contemporary Debate*. Oxford: Oxford University Press.

Bigelow, J., B. Ellis, and C. Lierse. 1992. The World as One of Kind: Natural Necessity and Laws of Nature. *British Journal for the Philosophy of Science* 43: 371–388.

Bird, A. 2007. *Nature's Metaphysics. Laws and Properties*. Oxford: Clarendon Press.

Bird, A., and Tobin, E. 2023. Natural Kinds. In *The Stanford Encyclopedia of Philosophy* (Spring 2023 edition), eds. E. Zalta, and U. Nodelman. https://plato.stanford.edu/archives/spr2023/entries/natural-kinds/

Blondeau, J., and M. Ghins. 2012. Is There an Intrinsic Criterion for Causal Lawlike Statements? *International Studies in the Philosophy of Science* 26: 381–401.

Boantza, V. 2007. Collecting Airs and Ideas: Priestley's Style of Experimental Reasoning. *Studies in History and Philosophy of Science* 38: 506–522.

Boyd, R. 1981. Scientific Realism and Naturalistic Epistemology. In *PSA 1980, Vol. 2. East Lansing*, ed. P.D. Asquith and T. Nickles, 613–662. MI: Philosophy of Science Association.

Boyle, R. 1991. In *Selected Philosophical Papers of Robert Boyle*, ed. M.A. Stewart. Cambridge: Hackett.

Bradley, F.H. 1963. *Appearance and Reality. A Metaphysical Essay*. Oxford: University Press.

Braithwaite, R. 1953. *Scientific Explanation*. Cambridge: Cambridge University Press.

Bresciani Filho, E., and I.M.L. D'Ottaviano. 2018. Basic Concepts of Systemics. In *Systems, Self-organization, and Information: An Interdisciplinary Perspective*, ed. A. Pereira Jr., W.A. Pickering, and R. Gudwin, 47–63. New York: Routledge.

Brewer, B. 2011. *Perception and Its Objects*. Oxford: Oxford University Press.

Bricmont, J. 2016. *Making Sense of Quantum Mechanics*. Cham: Springer.

———. 2022. *Making Sense of Statistical Mechanics*. Cham: Springer.

Brown, H.R. 2022. Do Symmetries "Explain" Conservation Laws? The Modern Converse Noether Theorem vs Pragmatism. In *The Physics and Philosophy of Noether's Theorems*, ed. J. Read, B. Roberts, and N. The, 144–167. Cambridge: Cambridge University Press.

Brown, H.R., and D. Wallace. 2005. Solving the Measurement Problem: de Broglie-Bohm Loses out to Everett. *Foundations of Physics* 35: 517–540.

Buber, M. 2010. *I and Thou*. Connecticut: Martino Fine Books.

Bueno, O. 2010. Models and Scientific Representations. In *New Waves in Philosophy of Science*, ed. P. Magnus and J. Busch, 94–111. Basingstoke: Palgrave Macmillan.

———. 2011. Scientific Representation: A Long Journey from Pragmatics to Pragmatics. *Metascience* 20: 423–428.

Carroll, J.W. 1990. The Humean Tradition. *The Philosophical Review* 99: 185–219.

———. 1994. *Laws of Nature*. Cambridge: Cambridge University Press.

Cartwright, N. 1983. *How the Laws of Physics Lie*. Oxford: Clarendon Press.

———. 1989. *Nature's Capacities and their Measurements*. Oxford: Oxford University Press.

Chakravartty, A. 2007. *A Metaphysics for Scientific Realism. Knowing the Unobservable*. Cambridge: Cambridge University Press.

———. 2010. Informational Versus Functional Theories of Representation. *Synthese* 172: 197–213.

Clarke, S., and T. Lyons, eds. 2002. *Recent Themes in the Philosophy of Science: Scientific Realism and Commonsense. Australasian Studies in History and Philosophy of Science Vol. 17*. Dordrecht: Kluwer.

Contessa, G. 2007. Scientific Representation, Interpretation, and Surrogative Reasoning. *Philosophy of Science* 74: 48–68.

Cordero, A. 2017. Retention, Truth-Content and Selective Realism. In *Scientific Realism: Objectivity and Truth in Science*, ed. E. Agazzi, 245–256. Cham: Springer.

Cushing, J.T., et al., eds. 1996. *Bohmian Mechanics and Quantum Theory: An Appraisal*. Boston Studies in the Philosophy and History of Science 184. Dordrecht: Springer.

Da Costa, N., and S. French. 2003. *Science and Partial Truth. A Unitary Approach to Models and Scientific Reasoning*. Oxford: Oxford University Press.

Daly, Chr. 2005. So, where is the Explanation? In *Truthmakers. The Contemporary Debate*, ed. H. Beebee and J. Dodd, 85–104. Oxford: Oxford University Press.

de Broglie, L. 1945. *La réalité des molécules et l'œuvre de Jean Perrin*. Paris: Gauthier-Villars. http://www.academie-sciences.fr/pdf/dossiers/Perrin/Perrin_pdf/Perrin_eloge.pdf.

Descartes, R. 1982. *Principles of Philosophy*. Trans. by V. Miller, and R. Miller. Dordrecht: Kluwer.

Devitt, M. 1997. *Realism and Truth*. Oxford/Princeton: Basil Blackwell/Princeton University Press.

———. 2020. An Ignored Argument for Scientific Realism. *Filozofia Nauki* 28: 5–24.

Dorato, M. 2005. *The Software of the Universe. An Introduction to the History and Philosophy of Laws of Nature*. Aldershot: Ashgate.

———. 2011. Mathematical Biology and the Existence of Biological Laws. In *Probabilities, Laws and Structure*, ed. D. Dieks et al. Dordrecht: Springer.

Douven, I. 2021. Abduction. In *The Stanford Encyclopedia of Philosophy* (Summer 2021 edition), ed. E. Zalta. https://plato.stanford.edu/archives/sum2021/entries/abduction/

Dowe, Ph. 2000. *Physical Causation*. Cambridge: Cambridge University Press.

Dretske, F.I. 1977. Laws of Nature. *Philosophy of Science* 44: 248–268. Reprinted in *Readings on Laws of Nature* ed. J. W. Carroll. 2004, 16–37. Pittsburgh: University of Pittsburgh Press.

Elder, C. 1994. Laws, Natures and Contingent Necessities. *Philosophy and Phenomenological Research* 54: 649–667.

Ellis, B. 2001. *Scientific Essentialism*. Cambridge: Cambridge University Press.

———. 2002. *The Philosophy of Nature. A Guide to the New Essentialism*. Chesham: Acumen.

———. 2008. Essentialism and Natural Kinds. In *The Routledge Companion to Philosophy of Science*, ed. S. Psillos and M. Curd, 139–148. New York: Routledge.

———. 2010. Causal Powers and Categorical Properties. In *The Metaphysics of Powers: their Grounding and their Manifestation*, ed. A. Marmodoro, 133–142. New York: Routledge.

———. 2012. The Categorical Dimensions of the Causal Powers. In *Properties, Powers and Structures. Issues in the Metaphysics of Realism*, ed. A. Bird, B. Ellis, and H. Sankey, 11–26. London: Routledge.

Ellis, B., and C. Lierse. 1994. Dispositional Essentialism. *Australasian Journal of Philosophy* 72: 27–45.

Feyerabend, P. 1958. An Attempt at a Realistic Interpretation of Experience. *Proceedings of the Aristotelian Society* 58: 143.

Frege, G. 1984. On Sense and Reference. In *Collected Papers on Mathematics, Logic, and Philosophy*, ed. B. McGuinness. Oxford: Blackwell.

Frigg, G. 2010. Models and Fiction. *Synthese* 172: 251–268.

Galileo. 1957. *Discoveries and Opinions of Galileo*. Trans. by S. Drake. New York: Anchor Books.

Ghins, M. 1990. *L'inertie et l'espace-temps absolu de Newton à Einstein. Une analyse philosophique*. Bruxelles: Académie Royale de Belgique. Classe des lettres. Tome LXIX/2.

———. 1992. Scientific Realism and Invariance. In *Proceedings of the Third SOFIA Conference on Epistemology. Campinas. July 30 – August 1, 1990. Philosophical Issues (Vol. 2: Rationality in Epistemology)*, 249–262. California: Ridgeview.

———. 2000. «Empirical *versus* Theoretical Invariance and Truth» (Followed by a commentary by Bas van Fraassen). *Foundations of Physics* 30: 1643–1655.

———. 2002. Putnam's No-Miracle Argument: a Critique. In *Recent Themes in the Philosophy of Science: Scientific Realism and Commonsense*, ed. S. Clarke and T. Lyons, 121–138. Dordrecht: Kluwer.

———. 2003. Les atomes et l'espace absolu: les raisons et la nature de l'antiréalisme de Mach. *Philosophia Scientiae* 7: 3–22.

———. 2005. Can Common Sense Realism be Extended to Theoretical Physics? *Logic Journal of the IGPL (Oxford UP)* 13: 95–111.

———. 2007. Laws of Nature: Do we Need a Metaphysics? *Fifth Principia International Symposium. Principia* 11: 127–114.
———. 2010. Bas van Fraassen on Scientific Representation. *Analysis* 70: 524–536.
———. 2013a. *Uma Introdução à Filosofia da Natureza. Representação, realismo e leis científicas*. Curitiba: Editora da Universidade Federal do Paraná.
———. 2013b. "Semirealism, Concrete Structures and Theory Change", Participation in a book symposium on *A Metaphysics for Scientific Realism. Knowing the Unobservable* by Anjan Chakravartty, *Erkenntnis* 78, 19–27.
———. 2014. Causal Powers as Metaphysical Grounds for Laws of Nature. In *Proceedings of the AIPS Conference in Široki Brijeg (Bosnia-Herzegovina), 24-27 July 2013*. Special issue of *Epistemologia*, ed. E. Agazzi. Milano: FrancoAngeli.
———. 2017. Defending Scientific Realism without Inference to the Best Explanation. *Axiomathes* 27: 635–651.
———. 2018. Are de Broglie and Bohm right? *Metascience* 27: 91–94.
———. 2019. Mechanistic Explanation: An Extension and Defence. In *Mechanistic Explanations in Physics and Beyond*, European Studies in the Philosophy of Science 11, ed. B. Falkenburg and G. Schiemann, 93–110. Cham: Springer.
Giere, R. 1988. *Explaining Science: A Cognitive Approach*. Chicago: University of Chicago Press.
———. 1999. *Science Without Laws*. Chicago: University of Chicago Press.
Glennan, S.S. 1996. Mechanisms and the Nature of Causation. *Erkenntnis* 44: 49–71.
Glennan, S. 2000. Rethinking Mechanistic Explanation. *Philosophy of Science* 69 (S3): S342–S353.
———. 2019. *The New Mechanical Philosophy*. Oxford: Oxford University Press.
Goodman, N. 1955. *Fact, Fiction and Forecast*. Indianapolis: Bobbs-Merrill.
———. 1976. *Languages of Art: An Approach to a Theory of Symbols*. Indianapolis: Hackett Publishing Company.
Hacking, I. 1983. *Representing and Intervening. Introductory Topics in the Philosophy of Natural Science*. Cambridge: Cambridge University Press.
Harré, R., and E. Madden. 1975. *Causal Powers: A Theory of Natural Necessity*. Oxford: Blackwell.
Hayek, A. 2019. Interpretations of Probability. In *The Stanford Encyclopedia of Philosophy* (Fall 2019 edition), ed. E. Zalta. https://plato.stanford.edu/archives/fall2019/entries/probability-interpret
Heil, J. 2003. *From an Ontological Point of View*. Oxford: Oxford University Press.
Hempel, C., and P. Oppenheim. 1948. Studies in the Logic of Explanation. *Philosophy of Science* 15: 135–175. Reprinted with a postscript in Hempel, C. (1965), *Aspects of Scientific Explanation and other Essays in the Philosophy of Science*, 245–295. New York: The Free Press.
Hesse, M. 1970. *Models and Analogies in Science*. Oxford: Oxford University Press.
Hume, D. 1978. In *A Treatise of Human Nature*, ed. L.A. Selby-Bigge and P.H. Nidditch. Oxford: Clarendon Press.
Ioannidis, S., V. Livanios, and S. Psillos. 2021. No Laws and (Thin) Powers in, No (Governing) Laws out. *European Journal for Philosophy of Science* 11: 1–26.
Jacobs, J.D. 2010. A Powers Theory of Modality: or How I Learned to Stop Worrying and Reject Possible Worlds. *Philosophical Studies* 151: 227–248.
———., ed. 2017. *Causal Powers*. Oxford: Oxford University Press.
Johnston, M. 1992. How to Speak of the Colours. *Philosophical Studies* 68: 221–263.
Kistler, M. 2006. *Causation and Laws of Nature*. London: Routledge.
Kitcher, Philip S. 1993. *The Advancement of Science*. Oxford: Oxford University Press.
———. 2001. Real Realism: The Galilean Strategy. *The Philosophical Review* 110: 151–194.
Kripke, S. 1980. *Naming and Necessity*. Cambridge, MA: Harvard University Press.
Kuhn, T.S. 1970. *The Structure of Scientific Revolutions, second edition, with postscript*. Chicago: University of Chicago Press.

Lakatos, I. 1978. In *The Methodology of Scientific Research Programmes. Philosophical Papers. Volume 1*, ed. J. Worrall and G. Currie. Cambridge: Cambridge University Press.
Lange, M. 2000. *Natural Laws in Scientific Practice*. Oxford: Oxford University Press.
Laplace, P. S. 1995. *Philosophical Essay on Probabilities*. Trans. by A. I. Dale. Dordrecht: Springer.
Laudan, L. 1981. A Confutation of Convergent Realism. *Philosophy of Science* 48: 19–49.
Laudan, L., and J. Leplin. 1991. Empirical Equivalence and Underdetermination. *Journal of Philosophy* 88: 449–472.
Leplin, J. 1997. *A Novel Defense of Scientific Realism*. Oxford: Oxford University Press.
Levin, M. 1979. On Theory-Change and Meaning-Change. *Philosophy of Science* 46: 407–424.
———. 1984. What Kind of Explanation is Truth? In *Scientific Realism*, ed. J. Leplin, 124–139. Berkeley: University of California Press.
Lewis, D.H. 1994. *We, the Navigators. The Ancient Art of Landfinding in the Pacific*, ed. Sir Derek Oulton. Honolulu: University of Hawaii Press.
Lewis, D.K. 1973. *Counterfactuals*. Cambridge, MA: Harvard University Press.
———. 1983. New Work for a Theory of Universals. *Australasian Journal of Philosophy* 61: 343–377.
———. 1986a. *On the Plurality of Worlds*. Oxford: Blackwell.
———. 1986b. *Philosophical Papers II*. Oxford: Oxford University Press.
———. 1997. Finkish Dispositions. *The Philosophical Quarterly* 47: 143–158.
———. 2008. Ramseyan Humility. In *Conceptual Analysis and Philosophical Naturalism*, ed. D. Braddon-Mitchell and D.R. Nola, 203–222. Cambridge, MA: MIT Press.
Lipton, P. 2004. *Inference to the Best Explanation*. 2nd ed. London: Routledge.
Livanios, V. 2010. Symmetries, Dispositions and Essences. *Philosophical Studies* 148: 295–305.
———. 2017. *Science in Metaphysics: Exploring the Metaphysics of Properties and Laws*. Cham: Palgrave Macmillan.
Locke, J. 1689. In *Essay Concerning Human Understanding*, ed. R. Woolhouse. London: Penguin Classics.
Lowe, E.J. 2002. *A Survey of Metaphysics*. Oxford: Oxford University Press.
Lyons, T. 2002. Scientific Realism and the Pessimistic Meta-Modus Tollens. In *Recent Themes in the Philosophy of Science. Australasian Studies in History and Philosophy of Science*, ed. S. Clarke and T. Lyons, 63–90. Dordrecht: Kluwer.
Machamer, P., L. Darden, and C. Craver. 2000. Thinking about Mechanisms. *Philosophy of Science* 67: 1–25.
Magnus, P.D. 2003. Success, Truth and the Galilean Strategy. *British Journal for the Philosophy of Science* 54: 465–474.
Marconi, D. 2007. *Per la verità. Relativismo e filosofia*. Turin: Einaudi.
———. 2009. Being and Being Called: Paradigm Case Arguments and Natural Kind Words. *The Journal of Philosophy* 106: 113–136.
Martin, C.B. 1994. Dispositions and Conditionals. *The Philosophical Quarterly* 44: 1–8.
Maxwell, G. 1962. The Ontological Status of Theoretical Entities. In *Scientific Explanation, Space and Time*. Minnesota Studies in the Philosophy of Science. Vol. 3, ed. H. Feigl and G. Maxwell. Minneapolis: University of Minnesota Press.
Meyer, A. 1995. *Oceanic Art – Ozeanische Kunst – Art océanien*. Cologne: Könemann.
Mill, J.S. 1843. *A System of Logic: Ratiocinative and Inductive*. London: Parker.
Molnar, G. 2003. *Powers*. Oxford: Oxford University Press.
Mozel, P. 1995. The Eclipse of Nicias. *Journal of the Royal Society of Canada* 89: 11–17.
Mumford, S. 1998. *Dispositions*. Oxford: Clarendon Press.
———. 2004. *Laws in Nature*. Abingdon: Routledge.
Musgrave, A. 1988. The Ultimate Argument for Scientific Realism. In *Realism and Relativism in Science*, ed. R. Nola, 229–252. Dordrecht: Kluwer.
———. 2017. Strict Empiricism Versus Explanation in Science. In *Varieties of Scientific Realism*, ed. E. Agazzi, 71–93. Cham: Springer.

Newton, I. 1961. In *The Correspondence of Isaac Newton, Vol. 3*, ed. H.W. Turnbull. Cambridge: Cambridge University Press.

———. 1971. *Sir Isaac Newton's Mathematical Principles of Natural Philosophy and His System of the World*. Trans. by A. Motte, rev. by F. Cajori. Berkely: University of California Press.

Nola, R. 2008. The Optimistic Meta-Induction and Ontological Continuity: the Case of the Electron. In *Rethinking Scientific Change and Theory Comparison: Stabilities, Ruptures, Incommensurabilities*, ed. L. Soler, H. Sankey, and P. Hoyningen-Huene, 159–202. Dordrecht: Springer.

Nye, M.J. 1972. *Molecular Reality. A Perspective on the Scientific Work of Jean Perrin*. London: Mac Donald.

Peirce, C.S. 1931. In *Collected Papers of Charles Sanders Peirce*, ed. C. Hartshorne and P. Weiss. Cambridge, MA: Harvard University Press.

Pero, F., and M. Suárez. 2016. Varieties of Misrepresentation and Homomorphism. *European Journal for Philosophy of Science* 6: 71–90.

Perrin, J. 1903. *Traité de chimie physique. Les principes*. Paris: Gauthier-Villars. https://iris.univ-lille.fr/pdfpreview/bitstream/handle/1908/3160/43122.pdf?sequence=1

———. 1909a. Mouvement brownien et grandeurs moléculaires. *Radium* 6: 353–360. https://hal.archives-ouvertes.fr/jpa-00242381/document.

———. 1909b. Mouvement brownien et réalité moléculaire. *Annales de chimie et de physique* 18: 5–114. http://hermes.ffn.ub.es/luisnavarro/nuevo_maletin/Perrin_1909.pdf.

———. 1910. *Brownian Movement and Molecular Reality*. Trans. of Perrin (1909b) by F. Soddy. London: Taylor and Francis. https://archive.org/details/brownianmovement00perr/page/n5/mode/2up?view=theater

———. 1913. *Les atomes*. Paris: Alcan.

———. 1916. *Atoms*. Trans. by D. L. Hammick of Perrin (1913). London: Constable & Company Ltd. https://archive.org/details/atomsper00perruoft/page/n11/mode/1up?ref=ol&view=theater

———. 1923. *Notice sur les travaux scientifiques de Jean Perrin*. Toulouse: Edouard Privat.

Plato. 1973. In *The Collected dialogues*, ed. E. Hamilton and H. Cairns. Princeton: Princeton University Press.

Portides, D. 2018. Idealization and Abstraction in Scientific Modelling. *Synthese* 198: 5873–5895.

Potters, J. 2019. Heuristics versus Norms: On the Relativistic Responses to the Kaufmann Experiments. *Studies in History and Philosophy of Modern Physics* 66: 69–89.

Prigogine, I., and I. Stengers. 1997. *The End of Certainty. Time, Chaos, and the New Laws of Nature*. London: The Free press.

Psillos, S. 1999. *Scientific Realism. How Science Tracks Truth*. London: Routledge.

———. 2002. *Causation and Explanation*. Montreal: McGill-Queen's University Press.

———. 2006. What Do Powers Do When They Are Not Manifested? *Philosophy and Phenomenological Research* LXXII: 157–176.

———. 2011a. Making Contact with Molecules: On Perrin and Achinstein. In *Philosophy of Science Matters: The Philosophy of Peter Achinstein*, ed. G.J. Morgan, 177–190. Oxford: Oxford University Press.

———. 2011b. Moving Molecules Above the Scientific Horizon: On Perrin's Case for Realism. *Journal for General Philosophy of Science* 42: 339–363.

———. 2014a. The view from Within and the View from Above: Looking at van Fraassen's Perrin. In *Bas van Fraassen's Approach to Representation and Models in Science*, ed. W.J. Gonzalez, 143–166S. Dordrecht: Springer.

———. 2014b. Regularities, Natural Patterns and Laws of Nature. *Theoria* 79: 9–27.

Psillos, S., and M. Curd, eds. 2008. *The Routledge Companion to Philosophy of Science*. London: Routledge.

Putnam, H. 1970. Is Semantics Possible? In *Mind, Language, and Reality, Philosophical Papers, vol. 2.*, ed. H. Putnam, 139–152. Cambridge: Cambridge University Press.

———. 1973. Meaning and Reference. *Journal of Philosophy* 70: 699–711.

———. 1978. *Meaning and the Moral Sciences*. London: Routledge.

Quine, W.V.O. 1948. On What There Is. *Review of Metaphysics* 2: 21–38.
———. 1951. Two Dogmas of Empiricism. *The Philosophical Review* 60: 20–43.
———. 1969. Natural Kinds. In *Ontological Relativity and other Essays*, ed. Jaegwon Kim and Ernest Sosa, 114–138. New York: Columbia University Press.
Ramsey, F.P. 1928. Universals of Law and of Fact. In *F. P. Ramsey, Foundations: Essays in Philosophy, Logic, Mathematics and Economics*, ed. D.H. Mellor, 128–132. London: Routledge.
Reichenbach, H. 1947. *Elements of Symbolic Logic*. New York: Macmillan.
———. 1949. *The Theory of Probability*. Berkeley: University of California Press.
Reimer, M., and E. Michaelson. 2019. Reference. In *The Stanford Encyclopedia of Philosophy* (Summer 2022 edition), ed. E. Zalta. https://plato.stanford.edu/archives/sum2022/entries/reference/
Ross, D. 1971. *Aristotle*. London: Methuen.
Russell, B. 1905. On Denoting. *Mind* 14: 479–493.
———. 1911. Knowledge by Acquaintance and Knowledge by Description. *Proceedings of the Aristotelian Society* 11: 108–128.
———. 1948. *Human Knowledge. Its Scope and Limits*. London: Allen & Unwin.
Ruyant, Q. 2021. *Modal empiricism. Interpreting Science without Scientific Realism*. Synthese Library 440. Cham: Springer.
Ryle, G. 1963. *The Concept of Mind*. London: Hutchinson.
Salmon, W. 1984. *Scientific Explanation and the Causal Structure of the World*. Princeton: Princeton University Press.
Sankey, H. 2008. Scientific Method. In *The Routledge companion to philosophy of science*, ed. S. Psillos and M. Curd, 248–258. London: Routledge.
Schrenk, M. 2017. *Metaphysics of Science. A Systematic and Historical Introduction*. London: Routledge.
Sendlak, M. 2018. On Quantitative and Qualitative parsimony. *Metaphilosophy* 49: 153–166.
Shoemaker, S. 1980. Causality and Properties. In *Time and Cause: Essays Presented to Richard Taylor*, ed. P. van Inwagen, 109–135. Dordrecht: Springer.
Smart, J.J.C. 1963. *Philosophy and Scientific Realism*. London: Routledge.
Spott, E. 1883. *Bismarck: The Book of Mistakes*.
Stanford, K.P. 2006. *Exceeding Our Grasp: Science, History, and the Problem of Unconceived Alternatives*. Oxford: Oxford University Press.
Suárez, M. 2003. Scientific Representation: Against Similarity and Isomorphism. *International Studies in the Philosophy of Science* 17: 226–244.
Suppe, F. 1974. *The Structure of Scientific Theories*. Chicago: University of Illinois.
Suppes, P. 1967. What is a Scientific Theory? In *Philosophy of Science Today*, ed. S. Morgenbesser, 55–67. New York: Basic books.
———. 2002. *Representation and Invariance of Scientific Structures*. Stanford: CLSI.
Tarski, A. 1944. The Semantic Conception of Truth and the Foundations of Semantics. *Philosophy and Phenomenological Research* 4: 341–376.
Tooley, M. 1977. The Nature of Laws. *Canadian Journal of Philosophy* 7: 667–698. Reprinted in Readings on Laws of Nature ed. J. W. Carroll. 2004, 38–70. Pittsburgh: University of Pittsburgh Press.
van Fraassen, B. 1980. *The Scientific Image*. Oxford: Oxford University Press.
———. 1989. *Laws and Symmetry*. Oxford: Oxford University Press.
———. 1995. Against Naturalised Epistemology. In *On Quine: New essays*, ed. P. Leonardi and M. Santambrogio, 68–88. Cambridge: Cambridge University Press.
———. 2000. Michel Ghins on the Empirical Versus the Theoretical. *Foundations of Physics* 30: 1655–1661.
———. 2002. *The Empirical Stance*. New Haven: Yale University Press.
———. 2008. *Scientific Representation. Paradoxes of Perspective*. Oxford: Oxford University Press.

———. 2009. The Perils of Perrin, in the Hands of Philosophers. *Philosophical Studies* 143: 5–24.
———. 2010. Reply to Contessa, Ghins, Healey. *Analysis* 70: 547–556.
———. 2017. Misdirection and Misconception in the Scientific Realism Debates. In *Varieties of Scientific Realism*, ed. E. Agazzi, 95–108. Cham: Springer.
Vetter, B. 2015. *Potentiality. From Dispositions to Modality*. Oxford: Oxford University Press.
Vickers, P. 2013. A Confrontation of Convergent Realism. *Philosophy of Science* 80: 189–211.
von Mises, R. 1957. *Probability, Statistics and Truth*, Second revised English edition. London: Allen & Unwin.
Whittaker, E. 1973. *A History of the Theory of Aether and Electricity. Vol. I: The Classical Theories*. New York: Humanities Press.
Wilson, C. 1997. *The Invisible World. Early Modern Philosophy and the Invention of the Microscope*. Princeton: Princeton University Press.
Wittgenstein, L. 1971. *Tractatus Logico-Philosophicus*. Trans. by D. F. Pears, and B. F. McGuinness. London: Routledge & Kegan.

Name Index

M. Ghins, *Scientific Realism and Laws of Nature: A Metaphysics of Causal Powers*, Synthese Library 483, https://doi.org/10.1007/978-3-031-54227-5

Subject Index

M. Ghins, *Scientific Realism and Laws of Nature: A Metaphysics of Causal Powers*, Synthese Library 483, https://doi.org/10.1007/978-3-031-54227-5

MIX
Papier aus verantwortungsvollen Quellen
Paper from responsible sources
FSC® C105338

FSC
www.fsc.org

If you have any concerns about our products,
you can contact us on
ProductSafety@springernature.com

In case Publisher is established outside the EU,
the EU authorized representative is:
Springer Nature Customer Service Center GmbH
Europaplatz 3, 69115 Heidelberg, Germany

Printed by Libri Plureos GmbH
in Hamburg, Germany